Using and Abusing Science

Using and Abusing Science:

Science and Political Discourse from Burke's "French Revolution" to Obama's Science Fair

Edited by

Véronique Molinari and Cyril Besson

Cambridge Scholars Publishing

Using and Abusing Science: Science and Political Discourse from Burke's "French Revolution" to Obama's Science Fair

Edited by Véronique Molinari and Cyril Besson

This book first published 2016

Cambridge Scholars Publishing

Lady Stephenson Library, Newcastle upon Tyne, NE6 2PA, UK

British Library Cataloguing in Publication Data
A catalogue record for this book is available from the British Library

ISBN (10): 1-4438-8945-8
ISBN (13): 978-1-4438-8945-2

TABLE OF CONTENTS

Part III: The Rhetorical Power of Science

INTRODUCTION

Over the last two centuries, as politics has evolved from the status of "amateurship" to that of profession, political discourse, its practices—the use of rhetoric among others—and their validity, have been increasingly questioned. Politicians, as illustrated by the low turnouts that have recently characterized general elections and a general lack of interest in politics throughout Western countries, enjoy less than ever the trust of the electorate; their discourse, as underlined by Christian le Bart, is now often criticized for being both hollow and untrustworthy, "predictable, coded, to the point of being potentially fraudulent, less interesting than self-interested" (2003, 1, 3). Conversely, by evolving from the status of enlightened amateur to that of expert, the figure of the scientist has, over the centuries, gained credibility with the general public. Even though, by presenting it as a form of discourse essentially aimed at convincing, Lyotard, in 1979, challenged its traditional view as the expression of reality (*La condition postmoderne*), science continues to be held in high regard and to be believed to provide a reliable form of knowledge. Summoning science has thus often been a way, in everyday life, advertisement and the popular media, to lend authority to a discourse, and imply that one's claims are well-founded or beyond dispute. "The naming of some claim or line of reasoning or piece of research 'scientific'", Chalmers explains, is "intended to imply some kind of merit or special kind of reliability" (xix). That politicians should have occasionally been tempted to do the same and make up for the deficit of legitimacy of their discourse through the instrumentalisation of scientific arguments or participation in contemporaneous debates on scientific issues is, therefore, not surprising. The issue at stake in this volume is to examine how, and to what extent, this process may have been taking place in the past two centuries.

Defining political discourse is not a straightforward matter. A discourse may be considered as "political" because of its source (the professionals of politics), its contents (it refers to current issues faced by the government), its mode of transmission or its effects (its electoral impact for instance) (Le Bart, 1998, 6). At the same time, "the political" can be defined so broadly that any discourse may be considered as "political". The choice, in this volume, has been not to limit political discourse to the discourse of experts but to exclude, nevertheless, the "everyday discourse of politics" (Wilson, 411). Will be considered as political discourse, in this study, discourses concerned with formal or informal political contexts and political actors, aimed at achieving political goals and presenting a political case.

Back in 1946, George Orwell, referring to the "lack of precision" of writers and their inability to express properly what they meant, wrote:

> The writer either has a meaning and cannot express it, or he inadvertently says something else, or he is almost indifferent as to whether his words mean anything or not. This mixture of vagueness and sheer incompetence is the most marked characteristic of modern English prose, and especially of any kind of political writing. As soon as certain topics are raised, the concrete melts into the abstract and no one seems able to think of turns of speech that are not hackneyed: prose consists less and less of *words* chosen for the sake of their meaning, and more and more of *phrases* tacked together like the sections of a prefabricated hen-house. [...] In our time, political speech and writing are largely the defence of the indefensible. Things like the continuance of British rule in India, the Russian purges and deportations, the dropping of the atom bombs on Japan, can indeed be defended, but only by arguments which are too brutal for most people to face, and which do not square with the professed aims of the political parties. Thus political language has to consist largely of euphemism, question-begging and sheer cloudy vagueness (1969).

The issue of manipulation and the aim of such manipulation for political goals, which were at the core of Orwell's comments, have been recurrent in works on political discourse, but with perhaps different levels of emphasis and analysis (Wilson, 400). In Fairclough's view, for example, political discourse is criticized as a "form of social practice with a malign social purpose" (in Wilson, 401). Instrumentalizing science, selecting scientific theories, not on the basis of objective criteria but on the basis of how helpfully they will serve their (political) goal, seizing upon scientific models or ideas so as to nurture and support their arguments on the assumption that science cannot be questioned and therefore one's discourse cannot be questioned—all these certainly pertain to such strategies of manipulation.

It is undoubtedly true that this misappropriation of science is made easier by a certain vagueness that seems to rest in the concept's DNA, if we consider its etymological meaning, as *scientia* is in Latin no different from the notion of knowledge itself. Through time, the term's breadth narrowed to something akin to a "body of knowledge organized in a systematic manner" (dictionary definition) and in a more modern usage still, "science" came to refer to a way of pursuing knowledge, not only knowledge itself. In the extra-linguistic world, this evolution could be seen as indiscernible from a no less gradual division (one feels tempted to say, "break") between an abstract form of philosophy (increasingly removed from the material world), with its claim to a universal understanding of "it all" (almost *a priori*, in the mind of the thinker), and a branch more grounded in the scrutiny of phenomena, no less ambitious perhaps but planning to achieve its ultimate goal, the explanation of everything, through different means. The method used requested a systematic study of the nature and behaviour of the material and physical universe, based on observation, experiment, and measurement, putting forward the formulation of *laws* to describe these facts in general terms.[1] Thus was born, or at least formalised, natural philosophy, the basis for what we now understand to be "science." The 17th and 18th centuries, especially, were a turning point, in that they publicized knowledge as increasingly formulated in terms of these "laws of nature." Consequently, in the course of the 19th century, the word "science" became increasingly associated with the scientific *method* itself to study the natural world, leading to a growing specialization of various disciplines (including physics, chemistry, geology and biology) that shared an awareness that different domains had to be approached in different ways, making their respective objects more manageable. It is also in the 19th century that the term scientist was first coined and increasingly replaced the expression "natural philosopher," signalling the consummation of the differentiation between philosophy and science, and the eventual advent of an autonomous scientific field. The latter, however, is to this day far from being unified, despite attempts to bring these "innumerable and endlessly diverse topics" that compose science under one formula.[2]

Given the topic of this book, it is of course not "science" as such that is to be examined in the following pages, but rather the *representation* of science, an image that rarely has much to do with the real thing, its inner workings or ultimate goals. Nor will we be concerned with the so-called "hard" sciences only, although they will be prevalent in this volume, as the symbolic weight they carry undoubtedly bestows more rhetorical power to those who wield the "scientific argument" than other—less undisputed—disciplines. Other sciences discussed in the following pages will include the so-called "social" sciences, such as economic science (even though its epistemic status is sometimes contested), as well as a flurry of others, some so discreditable for the common image of science as to be labelled "pseudo-sciences",[3] such as eugenics or phrenology. Even though the object here is not to enter this controversy, one can note that in this segregation, the central question is that of objectivity. For all intents and purposes, science is seen as objective, even objectivity itself, which posits it in the common imagination as at odds with politics, typically associated, as has been said earlier, with ideology. It is true that the two fields do not share the same goals: politics is concerned with the state of society and, even for conservatives, with a project, yet unrealized, but presumably oriented towards the improvement of this society. In this respect, referring to science enables the practitioner of politics to anchor his or her "unsubstantiated" discourse in something non transient, a field that has to do with facts and the reality of how the world works. Science is perceived as capable of abstracting itself from the world in the name of observation, refraining itself from all value-judgments or ideological leanings, and constituting a community whose inner conflicts bear little resemblance to polemics as expressed in the more aggressive realm of human affairs, the political arena. Thus, for some time, decision-making—at least in the Western world—has regularly been resorting to statistical science as a tool for guiding action and policies, conferring to them the authority of rationality.

Science appears, then, as a powerful, all-purpose argument. The difference between science itself and the way it is used is the claim to omnipotence that is made in the case of its exploitation for political purposes. Seen from within, what characterizes science is precisely the emphasis it places on the limits of its authority: the founders of modern philosophy of science especially, including Sir Karl Popper and members of the Vienna Circle like Otto Neurath, saw it as part of their role to explain the authority of science in terms of what falls within its bounds and which objects, on the other hand, elude its scrutiny. The corollary is that not everything will be explained one day.

While the progress of science, its ideas and its findings, can be an object of study in themselves, this volume, by examining the use of scientific arguments in political discourse, rather seeks to add to its cultural history, understood as the way it relates to society and is accommodated by culture. This link between political discourse and science will be examined, in its various forms, over a time-span covering a little more than two centuries, starting in late 18th century Britain and ending in the United States of the early 21st century. If we choose to consider this connection as a gradient, then three stages can be distinguished, from an indirect use of science as imagery to its direct integration–and the exploitation of its rhetorical power–in political texts and speeches.

The first example of a use of scientific imagery to serve a political discourse in this volume is drawn from the Revolutionary years, a time, according to Charles Coulston Gillispie, "when the density of the intersections [between politics and science] increased to a degree that is characteristic of modern polity in general" (2004, 1). The "scientific revolution", which is usually held to have begun with Copernicus (1473-1543) and ended with Isaac Newton (1642-1727),[4] had seen a core transformation, which had begun in cosmology and astronomy and then shifted to physics. For some historians, these changes in "natural philosophy", considered as the precursor of "natural sciences" or empirical science, had brought about important transformations in both ontology and epistemology (or how Europeans justified their claims to knowledge) (Hatch). Science, which was now dominated by scientific societies and academies as centres of research and development, played a leading role in Enlightenment discourse and thought. Parallel to the political upheavals of the French Revolution, remarkable scientific achievements took place and Paris became a vital centre for new scientific thought. Yet, as shown by Sarah Peyroux, for Edmund Burke, British statesman and political philosopher, the French Revolution was drawing on perverted scientific principles, which he opposed to the British scientific tradition—the basic seventeenth-century Baconian and Newtonian empirical philosophical and scientific methodology. The link between this essentially inductive scientific method and Burke's political philosophy, as illustrated by the author, is in fact the underlying reason for his thunderous denunciation of the French Revolution with its—almost *ex nihilo*—tendency to reconstruct the whole of French politics and administration on the basis of abstract–and essential deductive—reasoning. Burke himself traces his lineage to Bacon and the British empirical tradition when he compares the political observer to the chemist waiting for his preparation to become stable before drawing conclusions from his lab-experiment. In *Reflections on the Revolution in France*, one of the best-known intellectual attacks against the French Revolution, which Burke published in 1790, science however does not only apply to the philosophical basis of its author's thought. It is also part and parcel, through scientific similes, of the 'imagery' of his vituperative linguistic expression, which makes Burke, in addition to a philosopher, a powerful and effective polemicist.

Polemics is not what comes to mind first when one thinks of the work of Robert Louis Stevenson; yet he, too, famously brought into relief the bad uses of science in what can be considered a cautionary tale, *The Strange Case of Dr. Jekyll and Mr. Hyde* (1886). However, the 19th century Scottish writer does more than propose a simplistic charge against science gone wrong. In the novella, as in other fictional texts, he uses the *figure* of the doctor to reflect, in both senses of the word, on that profession, the way it was structuring itself at the time, and the mechanisms of control it exerted on the practitioners as well as the patients. Stevenson's doctor is the living embodiment of the ambivalent views of what the transformations of the medical profession entailed not just for physicians, but also for society, in terms of the distribution of social power. As attested by Jekyll's literally split personality, the Scotsman seems to have been, like his contemporaries, of two minds on the matter, a fact further demonstrated if one considers that the figure of the doctor in his fiction is counterbalanced by that of the engineer in his biographical writing. If the former is a nexus of uncertainties and ambiguities, where the individual is at every step faced with the demands of his profession to conform to his class, the latter comes closer to a secular saint, "naturally" aligned with the orthodox view of his vocation. His civic virtues, more than his expertise (perhaps because the former presume the latter) justify, or at least *should* justify, in the context of work, absolute authority at all levels, from the political microcosm of interpersonal relationships to the macrocosm of the alliance of nations, as the international fame of Stevenson's father is taken to indicate in the eulogy written by his son. Through those figures of the middle class professional, what is discussed in Stevenson's work is the rightful social place, in this sense political, not so much of science as of the *practitioner* of science, and where that leaves the society he operates in. Thus, science is shown to have practical uses, not just in its order (theoretical science does appear in his work, but his focus is much more on applied science), but also in that its concrete *praxis* translates into social and political modes of organization that Stevenson does not necessarily challenge, but with which he at least defamiliarizes his reader.

Philosophy and literature can construe science as part of their discourse on politics and society, but they often do so somewhat superficially. Science in Burke's and Stevenson's work certainly feels more like a mirror they use to envisage themselves and their surroundings than a concern to deal with "the real thing." Other fields will claim to have a more concrete approach to science, and maintain that by virtue of employing methods seen as scientific, they are truly a form of applied science, in a pragmatic approach. In the wake of social Darwinism, and especially the work of Spencer, there happened a dramatic "migration of scientific concepts to the field of human governance and political discourse" (Leblond) and, one might add, to the social field. The proponents of this migration undoubtedly saw themselves as promoters of the "genuine article," though arguably, they were no less in a figurative relationship to science as such, than those acknowledging a looser connection. Perhaps an operative distinction between "science of" and "science for" should be made at this point, between a science that observes, conservative of the world *as it is*, and one aimed at change, for which understanding is the prelude to the modification of the world by human agents. Scientific status, and the symbolic capital thereof, is what the agents of the social and political realities exposed in this second part claim for themselves, and with these agents, science is no longer defined as observation of the world, but as *action*.

With Andrew Carnegie, we move from one figure of cultural and social excellence (the scientist) to another (the successful capitalist). As Christian Leblond establishes, the American industrialist/ philanthropist certainly tried to appropriate the symbolic capital and the authority bestowed by science, seemingly unaware of, or perhaps unwilling to consider, the many differences between his field and biology. Whether a *pro domo* argument from someone who "could easily be dismissed as a purely opportunistic predator whose sense of ethics was primitive" or one in total earnest, Carnegie's use of the survival-of-the-fittest scenario, "building on the scientific discourse of Darwin and its subsequent transposition to human societies by Herbert Spencer," is of paramount importance for a correct understanding of his principles of management, later so influential with others of his class, translating into a new form of human organization spurred by the corporate world.

As C. Leblond's chapter attests, the prevalence of science in society and political discourse is not a given; a rhetorical appropriation is necessary before the claim to "being scientific" can be turned into social force by those who want to be seen as apt to wield science for the good of society and even the human race. Contrary to what we might think today, the "scientific argument" has not been seen as valid from time immemorial. Its history can be traced to a number of fundamental clashes between visions of the modes of regulation of society; the narrative that holds continuous, if sometimes uneven, progress for the place of science in social and political discourse, is a retrospective form of mythmaking disproved by many instances where discord, not agreement, was the norm. As J. Mullins shows, such a nodal conflict is exemplified by the controversies surrounding the 1846 Freeman case in the United States–a country where the supposed dominance of science is sometimes naturalized as an extension of the very Enlightenment principles upon which the nation was originally founded. Not so when one looks closely at the arguments developed during the trial by a number of prominent political figures, J. Van Buren and W. Seward among them, denoting a partisan rift among the American political class. Their tacit views on the relationship between science and politics had to be openly and publicly assumed before the bar, compelling the participants in the conflict "to give clear articulations of their social views and epistemological assumptions": the Whigs' reform project relied on the authority of science while the Democrats' views construed expertise as a confiscation of true democracy and egalitarianism, testifying to their differing views about the definition of a good society and ultimately, the meaning of freedom. In the end, the outcome of the trial led to something of a tie, for "if John Van Buren and the Democrats secured the legal victory, William Seward and the Whigs arguably won a moral victory".

The same preoccupations lie at the core of journalist and editor Benjamin O. Flower's crusade for "medical freedom" in the 1910s, advocating for the constitutional (in both senses of the word) right of Americans to choose from a whole array of medical practices not limited to conventional, or "allopathic", medicine. He did so in the name of individual freedom, but also as a defense of the integrity of American society. Using Herbert Spencer's evolutionary rhetoric to support his vision of progress, but also turning to the science of his day, Flower disputed the authority of normative, "state-sponsored" science, as part of an effort to regenerate and modernize the country. Curing the nation entailed the purification of society from the artificial barriers imposed by a restrictive medical legislation (the main manifestation of the bureaucratic order, a form of "class tyranny") to ensure a return to the natural, free circulation of men and ideas so as to restore the essence of America. This is transparent whenever this struggle is presented as a reenactment of the key episodes of American history, be it the Revolutionary War (the National League for Medical Freedom represented the "spirit of 1776" and "patriots" had a duty to defend American principles) or the role played in the construction of America by slavery, when early twentieth-century America is presented as divided between "medical slave states" and "medical free states." This raised a troubling epistemological and political question: who naturally held both the intellectual and political authority to decide and impose who the legitimate representatives of "science" were? Although Flower's reasoning is imbued with "the Protestant tradition of an unmediated relationship between the individual and his spiritual, and here physical, welfare" (so that Flower's crusade sometimes feels like "the Reformation redux" in Marin-Lamellet's felicitous phrasing), it would be too easy to discard his endeavour on grounds of religious and libertarian zealotry. In criticizing scientists' hubris, he developed a sense of what

Karl Popper would later called falsifiability, the fact that the scientific character of a concept resides in its refutability, sharing with the Pragmatists the idea that knowledge was tentative and collective, as well as an anti-elitist belief.

This concern for science's role in the freedom and welfare of human beings can be extended to other creatures than just the political animal, as the debates around the notion of animal rights illustrate. As E. Dardenne explains, for philosopher and ethicist Peter Singer, science enlightens us as to the sentience of animals, and their place in the ecological balance. It thus enables us to "better implement the values we have," while those moral premises are beyond the field of science itself, which cannot dictate a moral stance. His final goal is not scientific, however, or even truly philosophical for that matter, but practical: in the end, in his work, science-based statements of fact are connected with reasons for acting, not with moral judgments. Considering "the rights of orangutans" is then, contrary to Renan's argument, not absurd,[5] but political, although unconventionally so. Coming up with guidelines for a fair treatment of animals through a collective agreement would reconfigure the function and scope of politics. If "there might not be anything so distinctive about humanity that humans should have moral status and nonhumans should not," the field could then be dislodged from an anthropocentric position to one where all living creatures are accounted for. The end result of this could, if we follow Rolston, overcome even pathocentrism and lead to "a biocentric position, extending the scope of human responsibility to all living things," the *polis* now encompassing the whole of *Gaia*. Science would be a tool in the advent of this new ethic, leading to a dramatically enlarged view of the political.

The third part of this book will be devoted to the rhetorical power of science. "As a general rule", Auerbach, Christoph and Lenormand remark in the very last chapter, "the uses of science in politics tend to be *instrumental*–they are a means to an end." This aim is the accumulation of symbolic capital, what Bourdieu refers to as "the acquisition of a reputation for competence and an image of respectability and honourability…" (2013, 285). "Such indeed is the respect paid to science", Clerk Maxwell pointed out in 1871, "that the most absurd opinions may become current, provided they are expressed in language, the sound of which recalls some well-known scientific phrase" (quoted in Homchick). Science, because of the authority that is attributed to it, has therefore often been seized upon to advance political arguments and give more weight to a discourse, be it to contribute to the construction of women's domestic roles in 19th century

England or, conversely, to justify women's claim to political emancipation in the early 20th century.

The first contribution thus illustrates how, through the rhetoric and authority of science, Victorian scientific and pseudo-scientific texts could contribute to reinforce existing conceptions of gender in terms of physicality and intellect, particularly by underscoring the traits of female inferiority and domesticity. The works of Gall, Fowler, Galton, Pearson, Chambers and Darwin, Julie Homchick shows, all reveal the presence of ideological notions of female inferiority and domesticity. Interestingly while authors in phrenology and eugenics—two more overtly socially and politically-driven scientific theories—use scientific prose to promote a "separate sphere" agenda, Chambers' and Darwin's works, which appear much more descriptive than prescriptive, are imbued with similar messages regarding gender norms. Presented as scientific, the realm of the domestic sphere and traits of femininity could then be normalized and sanitized as part of the natural order of the sexes.

The women's movement that emerged in the late 19th century in Britain, while not necessarily questioning the existence of natural, or essential, differences between men and women such as those emphasized in the previous chapter, increasingly pressed for equal opportunities and equal rights in the field of education, employment and politics. The campaign for women's suffrage, among these, was to prove a long and difficult process, which lasted for more than half a century and culminated in the famous suffragette campaign of the early 1910s. The fact that, at this particularly tense period, an increasing number of militants should have felt it useful to put forward eugenic arguments to justify their claims for women's participation to parliamentary politics is yet another example of an instrumentalisation of science for political aims. The eugenic movement, which flourished in Britain at the turn of the century, can in itself be considered as a re-appropriation of scientific ideas which, mixed with a set of social theories, were to be applied to social policies. In this respect, the Darwinist and utilitarian origins of its arguments and, later, claims of statistical evidence, by giving it an appearance of scientific and pragmatic authority, undoubtedly contributed to its popularity at a time when concern for the future of the British "race" ran high. Even though it is usually considered today as a pseudo-science, eugenics was then presented as "the science which deals with all influences that improve the inborn qualities of a race" and its methods as "scientific." Considering the emphasis it put on motherhood at the expense of women's emancipation and the limits it consequently sought to impose to women's access to higher education and political rights, that some suffragists should have adopted such discourse may seem for the least surprising. To understand this process, and to determine whether the adoption of eugenic arguments in suffrage rhetoric stemmed from genuine adherence to eugenic concerns or corresponded to a process of instrumentalisation, Véronique Molinari takes a closer look at eugenic arguments and, more particularly, at the points of convergence with the feminist movement which made this reappropriation process possible. Taking the political background into account, she also tries to determine to what extent this argument of a "eugenic vote", put forward in a suffragist publication, stemmed from genuine eugenic concerns or was an attempt to instrumentalize the increasingly popular eugenic discourse and follow the Liberal government's concerns for national efficiency.

If political actors can use science with a view of advancing a political agenda, they can also contribute to shed light on science by putting in the limelight scientific issues that are often isolated in the anonymous world of their experts. Politicians, in particular, thanks to their status, can use their position of authority to promote science. One such example is provided by Barack Obama who, during his two terms in office, entertained, as shown by Gregory Benedetti, a bilateral relation of cooperation with the world of science. Not only did the 44th President of the United States, through his speeches, seek to offer the image of a politician who was supportive of the scientific community and to draw an ideological contrast with the Republicans on scientific issues; he also promoted initiatives and policies that aimed at putting science back at the forefront of the political arena and projecting a vision of science based on pragmatism, not on ideology. The annual White House Science Fairs, launched by Obama himself in 2010 following a commitment made at the launch of his *Educate to Innovate* campaign to "move American students from the middle to the top of the pack in science and math achievement over the next decade" (The White House, Office of the Press Secretary 23 November 2009) provided the President with opportunities, through his speeches, to regularly put forward his vision of science. Reuniting, not only science and politics, but also science and the younger generations, the Democratic Party, as illustrated by Benedetti, has thus presented itself as an ally of the scientific sphere, not only for the sake of science, but, more importantly, to advance a political agenda, reconnecting science (often perceived as exclusive and elitist) with social and ethnic minorities.

While the previous chapters have essentially focused on the natural sciences, the last chapter will investigate the legitimising effects of scientific discourse in relation to economics, with a particular focus on the interactions between economic science and political discourse during the

1970s and 1980s and on the role played by the economic theories underlying neoliberalism and the Alternative Economic Strategy put forward by post-Keynesian economists in the respective success and failure of these political projects. To understand how, in the 1980s, many observers left and right, in the United Kingdom, came to accept Margaret Thatcher's argument that there was no alternative to free-market capitalism, Paul Auerbach, Gilles Christoph and Marc Lenormand focus on the issues of inflation and the trade unions as conceptualized by proponents of neoliberalism and the AES as well as on their political and electoral impact.

Cyril Besson & Véronique Molinari
Univ. Grenoble Alpes ILCEA4, F - 38040 Grenoble

Bibliography

Bourdieu, Pierre. 2013. *Distinction: A Social Critique of the Judgment of Taste* (translated from the French). London: Routledge. [1984]

—. 1991. *Language and Symbolic Power* (translated from the French). Cambridge: polity. [1982]

Boyd, R., P. Gasper & J. D. Trout (eds). 1991. *The Philosophy of Science*, Cambridge, Massachusetts: MIT Press.

Chalmers, Alan F. 1999. *What is this thing called Science?* (third edition). Indianapolis: Hackett Publishing Company, Inc.

Fairclough, Norman. 1989. *Language and Power*. London: Longman.

Gillispie, Charles Coulston. 1959. "Science in The French Revolution" in *Proceedings of the National Academy of Sciences of the United States of America*, Princeton University, May, vol. 45(5): 677–684.

—. 2004. *Science and Polity in France: The Revolutionary and Napoleonic Years*. Princeton: Princeton University Press.

Hatch, Robert A. "The Scientific Revolution, Definition – Concept – History" at http://users.clas.ufl.edu/ufhatch/pages/03-Sci-Rev/SCI-REV-Teaching/03sr-definition-concept.htm (accessed Sept. 28, 2015)

Henry, John. 2002. *The Scientific Revolution and the Origins of Modern Science*. London: Palgrave Macmillan. [1997]

Holly, W. 1989. "Credibility and Political Language", in Ruth Wodak (ed.), *Language Power and Ideology*. Amsterdam: John Benjamins, pp. 115-35.

Kuhn, Thomas S. 1962. *The Structure of Scientific Revolutions*, Chicago: The University of Chicago Press.

Latour, Bruno. 1989. *La science en action*. Paris: La Découverte.

Le Bart, Christian. 1998. *Le discours politique*. Paris: Hachette.
—. 2003. "L'analyse du discours politique : de la théorie des champs à la sociologie de la grandeur". *Mots. La ville, entre dire et faire*, 72 (July).
Orwell, George. 1946. *Politics and the English Language*, First published in *Horizon*, London, April. To be consulted at: http://www.orwell.ru/library/essays/politics/english/e_polit/
Raynaud, Dominique. 2015. *Scientific Controversies: A Socio-Historical Perspective on the Advancement of Science* (translated from the French). London: Transaction Publishers. [2003]
Reid, Charles. "Ethos, Pathos, and Logos: A Guide for Scientists." http://charlesreid1.com/wordpress/2013/01/ethos-pathos-and-logos-a-guide-for-scientists (accessed Aug. 14, 2015).
Shapiro, Michael J. 1981. *Language and Political understanding: the Politics of Discursive Practices*. New Haven: Yale University Press.
Torode, B. 1991. *Review of Fairclough, N., Language and Power*, *Discourse and Society* 2 (1), pp.121-2.
Wilson, John. 2001. "Political discourse", in Deborah Schiffrin, Deborah Tannen, Heidi E. Hamilton, *The Handbook of Discourse Analysis*, Oxford: Blackwell, pp. 398-414.

Notes

[1] See the chapter "The Scientific Method" in John Henry, *The Scientific Revolution and the Origins of Modern Science*, pp. 14-52.

[2] "These very different activities and disciplines all involve systematic and unbiased observations; the due examination of the record of these by trained minds leads to classification; from such classifications general rules or "laws" are deduced; these laws may be applied to further observations; failures and correspondence between new observations and accepted laws may result in alterations of the laws; and these alterations lead to yet further observations; and so on. This chain of activities is usually held to constitute the 'method' of science". *Encyclopaedia Britannica*, 1962, vol. 20, p. 114, col. 1.

[3] The central issue here is what Karl Popper identified as "the demarcation problem," that is, finding an operative criterion to distinguish between empirical science (for example Einstein's general theory of relativity) and pseudoscience (psychoanalysis for instance). Popper opted for the notion of "falsifiability" as a decisive factor for demarcation. To illustrate the way this appraisal works, one can say that Freud's theories could not be disproved because intrinsically, they virtually prohibited the test of their hypotheses, closing the system to refutability, while Einstein's theory, even though originally impossible to prove concretely, could always be tested and contested through the *verifiable predictions* it made. Popper's concept has been challenged, by Feyerabend most notably.

[4] For a discussion of this disputed term, see J. Henry, *op. cit.*, pp. 1-8.

[5] In a letter to Strauss, Renan famously disputed the views of a "Germanic" patriotism that in the nineteenth century predicated a justification of German expansionism on a genealogic argument, whereby an antecedent culture was authorized by right of ancestry to recover its territorial integrity, the loss of which was but a historical accident: "With that philosophy of history," he notoriously wrote, "the only legitimate right that will be left eventually, will be that of orangutans, assumed to have been unfairly disenfranchised by the treachery of the civilized." To be consulted at: http://classiques.uqac.ca/classiques/renan_ernest/qu_est_ce_une_nation/qu_est_ce_une_nation_texte.html (accessed Dec. 19, 2015).

PART I:

THE IMAGERY OF SCIENCE

Scientific Imagery in the Reflections on the Revolution in France (1790): Edmund Burke's Vindication of the British Scientific Tradition

Sarah Peyroux-Roblin

When Edmund Burke published his *Reflections on the Revolution in France* in 1790, this sharp denunciation of the French revolutionary movement came as a surprise to most of his contemporaries. Not only had Burke never been hostile to revolutions (in 1774-75, he had even written speeches to support the American Revolution), but in more recent years he had ardently fought against established privileges.[1] His *Reflections* put an end to Burke's reputation as a radical. His old friends of the Whig party, such as Charles James Fox, broke with him. And the new English Jacobins, such as Thomas Paine, turned him into their arch-enemy, so passionate was the debate triggered off in England by the *Reflections on the Revolution*.

The book is presented as a letter "intended to have been sent to a gentleman in Paris," but the targeted reader might be more British than French. This long letter falls into two parts: in the first part, Burke opposes Richard Price, an enthusiast of the Revolution, who had contended that the French Revolution was the heir of the British Glorious Revolution. In the second part, Burke deals more theoretically with the faults of the new French regime, which allows him to highlight *a contrario* the merits of the English system.

The *Reflections* have always been praised for their style. To most critics, Burke is the last representative of the great rhetorical tradition of the eighteenth century, in the manner of Bolingbroke. His style is fluid and conversational in accordance with the epistolary form he has chosen; it is also lively, varied, and florid almost. James Boswell, the famous biographer and friend of Samuel Johnson, said of him that he had a great facility with stylistic devices, that he "'picked them up like apples."[2] And images are profuse in the *Reflections'* fruit basket.

Burke borrows them from a whole range of gentleman's concerns: from painting and architecture, literature and drama, finance and trade. However, the most pervasive network of images is that of science: it covers the whole variety of eighteenth-century scientific endeavours, from physics to chemistry, from mathematics to natural history, testifying to the great familiarity of the author with scientific debates, in accordance with his early training at the dissenting academy of Ballitore in Ireland (Fuchs, 22-23).

Of course, those images are not a mere ornament in Burke's prose: they have a real rhetorical function, as they help him make his point more explicit, and convince his reader of the validity of his counter-revolutionary stance. This paper will thus examine how scientific images are picked up and used in the *Reflections*. It will contend that this imagery is central to Burke's ideological enterprise, which heavily relies on the British scientific tradition, and its praise of empiricism. In *The Reflections* scientific metaphors have a twofold function: satirical and epistemological. On the one hand, in order to cast the blame on the French revolutionary movement, Burke repeatedly depicts it as either partaking of unscientific pursuit, or of over-speculative tendencies, which may prove dangerously impractical. On the other hand, to oppose the French method, he suggests a more empirical approach to politics, which is directly inspired by Bacon's inductive reasoning; if scientists are not qualified to be statesmen, governing a country should nevertheless be seen as a scientific experiment.

The Satirical Use of Scientific Metaphors: Blaming the French Revolutionaries as Incompetent Scientists

Burke's rejection of the French Revolution is articulated around three types of scientific prejudice that his reader, as a true heir of the seventeenth-century scientific revolution, was very likely to share: a modern scorn for pseudo-scientific or unscientific pursuits like alchemy or quackery, a more distinctively British contempt for speculative science in the tradition of Swift, and eventually a sort of blind devotion to Sir Isaac Newton, the English national hero of science.

Since the seventeenth century and the days of the scientific revolution, alchemists and quacks had been openly scorned by the British enlightened public as unscientific people who professed a false knowledge out of irreligious madness or selfish interest. Capitalizing on this negative image, Burke, in *The Reflections*, repeatedly compares the French revolutionaries to alchemists or quacks in order to deride their policies as dubious and ominous practices.

In his long development on the French revenue, Burke explicitly draws a parallel between the new French leaders and the alchemists of the past: “Their fanatical confidence in the omnipotence of church plunder has induced these philosophers to overlook all care of the public estate, just as the dream of the philosopher's stone induces dupes, under the more plausible delusion of the hermetic art, to neglect all rational means of improving their fortunes” (*Reflections*, 230). The creation of paper money (the *assignats)* following the confiscation of the Church estates by the new State is indeed presented by Burke as the French discovery of the philosopher's stone, as made clear by the choice of the term “transmutation” to describe that process: “It consists in the means of drawing out at discretion portions of the confiscated lands for sale; and carrying on a process of continual transmutation of paper into land, and land into paper” (187). Not only does this imagery allow Burke to insist on the dangerous consequence of such a policy, the “volatilisation” of property, one of the pillars of civil society, but it also immediately casts discredit on the theoretical basis of the French monetary practice, which appears just as illusory as the hermetic art of alchemy.[3]

Burke's portrayal of the French “alchemists” is more alarming than amused: with the art of the caricaturist, he depicts the revolutionaries in their infernal laboratories, their alembics of hell “furiously boiling” (88). And if they are just as greedy as their ancestors, their mad quest for money may turn to more sacrilegious means: “...the project for coining into money the bells of the suppressed churches. This is their alchymy. There are some follies which baffle argument; which go beyond ridicule; and which incite no feeling in us but disgust; and therefore I say no more about it” (238). As they are blind, fanatical and irresponsible, those alchemists' experiments will inevitably prove destructive.

Revolutionary quacks may be just as harmful for the country. Indeed the lexical field of quackery pervades the text of *The Reflections,* as Burke readily assimilates the French revolutionaries to medical impostors who are trying to cure dysfunctions in the existing political system. For those humorous portraits, Burke was inspired by seventeenth-century French comedies, a literary tradition he was familiar with,[4] but he also elaborates on the parallel between the revolution as a political strategy and the quacks’ much-vaunted nostrums.

Burke does not deny that the existing French political body was impaired, and needed to be reformed but, according to him, the treatment chosen by the French revolutionaries was too harsh: it consisted in “taking periodical doses of mercury sublimate, and swallowing down repeated provocatives of cantharides” (60), in using “poisonous weeds” and “wild incantations" likely to "hack the aged parent in pieces" (93): under the guise of physicians, they are the real executioners of the State, the torturers of the monarch.[5]

In several cases, Burke more precisely identifies the "Rights of Man" to the new nostrum that revolutionary quacks sell to the population and may try to export, but Burke is confident that the British public will not be gullible, having been trained to more enlightened medical practices, and that the French Revolution will be kept at a distance: "Your affairs, in spite of us, are made a part of our interest; [...] If it be a panacea, we do not want it. We know the consequences of unnecessary physic. If it be a plague, it is such a plague that the precautions of the most severe quarantine ought to be established against it" (86). However, Burke acknowledges the persuasive power of the Jacobins' rhetoric, of their equalitarian philosophy, a nefarious opiate which deludes the people into oblivion, into a passive acceptation of their misery:

> They tell the people, to comfort them in the rags with which they have clothed them, that they are a nation of philosophers; and, sometimes, by all the arts of quackish parade, by show, tumult and bustle, sometimes by the alarms of plots and invasions, they attempt to drown the cries of indigence, and to divert the eyes of the observer from the ruin and wretchedness of the state. (130)

The revolutionary leaders' oratorical skills are not so different from the quacks', addressing the crowd from platforms, sometimes raising the enthusiasm of the audience with an unbridled eloquence, sometimes manoeuvring to divert its attention and deceive it. This satirical parallel, implicitly drawn by Burke in *The Reflections,* was to be pictorially developed by James Gillray a few years later. In the context of the counter-revolutionary campaign, Gillray published *Copenhagen House* in November 1795[6], four days after the mass outdoor meeting of the London Corresponding Society (the English revolutionary society). This cartoon, inspired by seventeenth-century Dutch painting (in particular by Gerrit Dou's 1652 *Quack*[7]), portrays English Jacobins as quacks during a fair. As with Burke, the cartoonist's aim is to arouse the suspicion of the British public, to unveil the political imposture.

Besides the rejection of alchemy and quackery as unscientific knowledge, the scientific revolution of the seventeenth century laid the stress on experimentations and empiricism, and cast the blame on abstract scientific speculation. The general distrust among the British public for a purely theoretical approach is a scientific prejudice Burke often relies on in his metaphorical choices.

This outlook had been fuelled, to some extent, by Jonathan Swift's *Gulliver's Travels* (part 3), which features the fictional lands of Laputa and Balnibarbi. In those places, speculative science rules supreme: the inhabitants, with one of their eyes turned inward, and the other up to the zenith, are completely wrapped up in their speculations and blind to practical concerns. At the end of the day, with their lack of interpersonal exchange and ill-shaped material world, the two scientific utopias turn out to be hellish places to live in. No wonder that when he compared the newly-organized French country to those famous literary creations Burke was sure to win the reader's intertextual consent: "From the general aspect of the country one would conclude that it had been for some time past under the special direction of the learned academicians of Laputa and Balnibarbi" (129). The English reader would rather flee away from such a world, than accept these incompetent rulers.

Even before the eighteenth century, the utopian genre cast doubts on scientifically-constructed social systems. In the description of the new territorial organization decided by the National Assembly (*Reflections,* part two) Burke's style is reminiscent of Thomas More's narrator, Raphael Hythloday, presenting the geographical and political organization of the island of Utopia; behind the apparent delight in precise figures and new terminology, the authors ironically point at the absurdity of such geometrically-planned and arithmetically-organized systems:

> ...they divide the area of their country into eighty-three pieces, regularly square, of eighteen leagues by eighteen. These large divisions are called *Departments*. These they portion, proceeding by square measurement, into seventeen hundred and twenty districts, called *Communes*. These again they subdivide, still proceeding by square measurement, into smaller districts called *Cantons*, making in all 6400. (169)[8]

State surveyors have a leading role in the new French utopia: they have imposed their abstract geometrical rules on the country, dividing it along exact measurement and calculation, so much so that Burke greets an organization "made on the system of Empedocles and Buffon, and not upon any politic principle" (170). Hence, he cannot but predict the failure of a territorial division that ignores demographic, economic and cultural realities. To sum up his argument, in a Latin formula that sounds like a tribute to the humanist tradition: "*Hominem non sapient*" (178); Burke denounces a de-humanizing method that neglects human concerns.

With its emphasis on rationality (against the intellectual erring of alchemy and quackery) and on experimentalism (against the dangerous abstractions of speculation), the scientific revolution of the seventeenth century was epitomized by the works of Sir Isaac Newton. The poets of the first half of the eighteenth century largely contributed to the popularization and quasi-veneration of the English scientist,[9] so much so that, by the middle of the century, debasing his thought had become truly repugnant to the British public. In the *Reflections,* some scientific images are clearly related to Isaac Newton's theories: Burke's satirical strategy here consists in showing how the French revolutionaries are guilty of misunderstanding or misapplying Newton's scientific heritage. For instance, in the midst of his diatribe against the new territorial divisions, Burke makes the following remark:

> Is every land-mark of the country to be done away in favour of a geometrical and arithmetical constitution? (...) Are all orders, ranks and distinctions to be confounded, that out of universal anarchy, joined to national bankruptcy, three or four thousand democracies should be formed into eighty-three, and that they may all, by some sort of unknown attractive power, be organized into one? (52)

This "unknown attractive power" sounds distinctly Newtonian to the reader, but the French revolutionaries' attempt to apply the theory of universal attraction to their new social and geographical entities seems a desperate, pathetic, attempt at sorting out order out of chaos: obviously, for Burke, the potent optimism of the early eighteenth century, in the Thomsonian manner, cannot hold in a revolutionary context.

Burke's allusion to the other theoretical achievements of the famous English physicist is even more far-reaching:

> These metaphysic rights entering into common life, like rays of light which pierce into a dense medium, are, by the laws of nature, refracted from their straight line. Indeed in the gross and complicated mass of human passions and concerns, the primitive rights of men undergo such a variety of refractions and reflections, that it becomes absurd to talk of them as if they continued in the simplicity of their original direction. The nature of man is intricate. (59)

The notion of "light" in the philosophical context of the age opens onto a rich polysemy: it is the whole philosophy of the French Enlightenment that is here attacked, insofar as its abstract and rationalist dimension (its "metaphysic" tone to quote Burke's word) has no grasp on human nature: according to the British philosopher, the political scientists of the new French regime have failed to identify the very special nature of the medium they were working on, a medium which is of infinite variety and far from neutral. To Burke's mind, human nature cannot be reduced to mathematical principles, and this criticism forms the basis of his redefinition of political action and philosophy.

The Epistemological Use of Scientific Metaphors: Propounding a New Methodology for Scientific Politics

Alongside his satirical remarks, Burke adopts a more constructive approach when he analyses the proper links between politics and science: after excluding the abstract scientist from the political arena, he advocates an experimental approach for the political thinker, and announces a sort of epistemological revolution for political philosophy, on the Baconian model.

Even though some of his assertions may sound dangerously anti-intellectual,[10] Burke is not hostile to science and scientists. When he wrote his pamphlet, during the first stages of the Revolution, Burke knew that some leading scientists were among the key figures of the new regime: Jean-Sylvain Bailly, an astronomer, had been elected president of the National Assembly in June 1789; and the Marquis of Condorcet, a famous mathematician and philosopher, was secretary of the same assembly and very influential during legislative debates. In the *Reflections*, Burke does not hide his admiration for such men, addressing them as "men of shining talents" (38). But he laments that they have been deluded into actively taking part in the new institutions. To his mind, scientists should not interfere in the political life of the country. Be they the most brilliant minds of their age or the most obscure physicians of some French province called to sit on the benches of the new Assembly, they have not, according to Burke, received a proper training to rule a country: "the sides of sick beds are not the academies for forming statesmen and legislators" (41) (laboratories would probably not have seemed more appropriate to him). Burke's condemnation of the scientists' political involvement reflects his belief in the specialization of society, his rejection of *amateur* politics. But he may also want to preserve science just as much as politics since, for him, science is likely to be the first victim of the intrusion.

According to Burke, when scientists become statesmen, science will necessarily be degraded by the mob. This elitist vision of science is contained in a famous quotation that was to infuriate all the supporters of the *Rights of Man*: "Happy if learning, not debauched by ambition, had been satisfied to continue the instructor, and not aspired to be the master! Along with its natural protectors and guardians, learning will be cast into the mire, and trodden down under the hoofs of a swinish multitude" (76).

Scientists should rather remain in their ivory tower than run the risk to have their thoughts and theories disfigured by ignorant people. In his essay, Burke makes it clear that the political leaders of the movement have turned scientific notions of abstraction and universality into mere instruments of their equalitarian rhetoric: science was a means to reach their political goals, not an end for them. One of the first measures testifies, to Burke's mind, to this lack of concern: the confiscation of the church estates, decided in November 1789. This decision amounts to a major financial loss for scientists who are hence deprived of a traditional source of patronage.[11] To him, the Revolution tolls the knell of French science.

Furthermore, Burke hints at the totalitarian potentialities of the scientific discourse and shows how it has been put into practice by the presumptuous leaders of the movement. Invested with the prestige and dignity of their scientific supporters, they have simplified scientific theories to the extreme, turning them into a sheer rule of numbers: "It is said that twenty-four millions ought to prevail over two hundred thousand. True, if the constitution of a kingdom be a problem of arithmetic. This sort of discourse does well enough with the lamp-post for its second: to men who may reason calmly, it is ridiculous" (49). Burke contests the hegemony of a blind, arbitrary, scientific discourse that leaves no room for discussion. His remark sounds like an early premonition of the violent excesses of the Terror, with the lamp-posts serving as gallows to get rid of the enemies of the new regime, a practice so vividly depicted by Gillray on his 1793 cartoon, *The Zenith of French Glory.*[12]

Science thus has to be protected from the rashness and opportunism of revolutionaries, scientific discourse should never become prescriptive in order to preserve individual liberties; however, scientific methods, insofar as they encourage prudence, may guide observations and prove valuable to political chroniclers and philosophers.

In a metaphor that runs through the first pages of the *Reflections,* Burke explains how he has himself adopted a scientific method of observation:

> When I see the spirit of liberty in action, I see a strong principle at work; and this, for a while, is all I can possibly know of it. The wild *gas*, the fixed air, is plainly broke loose: but we ought to suspend our judgment until the first effervescence is a little subsided, till the liquor is cleared, and until we see something deeper than the agitation of a troubled and frothy surface (6-7).

Echoing the interest in chemistry stimulated by the recent and simultaneous discoveries on gases (especially oxygen) by Joseph Priestley in England, and Laurent-Antoine de Lavoisier in France, Burke plays the chemist in this excerpt: like an experimenter in front of his apparatus, he does not yield to the hurry of immediate observations, but is careful to wait until the process is fully stabilized, until all the effects have become manifest. According to him, patience, circumspection, those essential qualities of the scientific observer, should also be cultivated by the historian dealing with contemporary events. And with such analogies, Burke announces a modern vision of history as a social science, of historians as social scientists.

This new vision is also perceptible in the vocabulary used by Burke, who often resorts to mathematical or biological concepts in his historical interpretations. For instance, he uses the concept of "inverse ratio" to try and depict the correlation between the force of public opinion and the degree of popular authority (91), and also successfully distinguishes between the 1789 French Revolution and the 1688 British Revolution thanks to scientific analogies. Using a mathematical metaphor, he analyses the English episode as "a temporary solution of continuity" in the strict order of hereditary succession (16), while the French events are presented as an explosion of the existing order which has let loose "the organic *moleculae* of a disbanded people" (20). The italics in the last quotation draw the reader's attention to a biological neologism: the Latin term has not yet entered the English language and stands out in Burke's prose. It stands out just like the numerous original quotations from ancient writers (Cicero or Tacitus) expected at the end of the eighteenth century in a political essay. With their two different origins, Latin words highlight Burke's peculiar position, in-between a classical and a scientific age. In any case, he belongs to an age when scientific concepts become helpful to the descriptive task of the political philosopher.

But the *Reflections* go beyond a mere chronicle of the French revolution: they open onto a real vindication of the grounds and methods of British political philosophy. As he very violently rejects the "abstract and geometrical" politics of the French revolutionaries, Burke advocates an epistemological revolution on the Baconian model.

Bacon, in his writings (particularly *Novum Organum,* 1620), had supported a scientific approach based on experiments and observations, from which theories could be built: this method, which insists on deriving theories, physical laws, from facts (and not the other way round) is called inductive reasoning and was one of the pillars of the seventeenth-century scientific revolution in Britain. It constituted the founding principle of the Royal Society (1660) whose spiritual father was Bacon.

Burke's definition of political philosophy is distinctly Baconian: "The science of constructing a commonwealth, or renovating it, or reforming it, is, like every other experimental science, not to be taught *a priori*" (58). Like his scientific counterparts, Burke extols the inductive approach of experimental science, ascending from facts to theory: French politics is rejected as the avatar of a corrupt scientific method, a deductive reasoning that sounds dangerously restrictive and prescriptive. In the reversal he advocates, Burke is sometimes provocative: for instance, when he celebrates the value of prejudices: "Many of our men of speculation, instead of exploding general prejudices, employ their sagacity to discover the latent wisdom which prevails in them" (84). Prejudices are the precise target of the rationalist philosophers of the Revolution, who discard all notions that have no other ground than tradition and conformism. But Burke adopts the opposite point of view: to his mind, the study of customs and prejudices, of morals and manners, seems necessary to delineate the fittest political regime for the country. Consequently, history becomes the golden book of the political philosopher, and jurisprudence is elevated to the rank of a new science.[13]

Throughout his essay, Burke's preoccupation is with the living world: more than once he compares the French theorists of the Revolution with natural historians who study dead animals ("stuffed birds", 83) or work with dried plants (a "*hortus siccus*", 11); all those parallels lay the stress on the deadly dimension of the Revolution. On the contrary, Burke insists that British politics have always preserved the living principle of the country: as Englishmen have generally preferred reforms to revolutions, they are readily compared to gardeners or agronomists who are wary to engraft acceptable scions onto an original plant ("upon that body and stock of inheritance we have taken care not to inoculate any scion alien to the nature of the original plant", 29), an "original plant" that might be a reference to the native tree of British politics, the old oak of English liberties. They are also likened to doctors of physic who tend to their patients with care and prudence ("man should approach to the faults of the state as to the wounds of a father, with pious awe and trembling solicitude", 93) and sadly acknowledge the necessity to administer "a bitter potion to a distempered state" (28). With all these metaphors, Burke relates English politics to applied modern sciences, which take into account the real world, and try to improve it, to make it healthier or more productive. In the context of the late eighteenth century, and of the emerging Industrial Revolution, this imagery even finds a new echo:

> A politician, to do great things, looks for a *power*, what our workmen call a *purchase;* and if he finds that power, in politics as in mechanics, he cannot be at a loss to apply it. [...] To destroy any power, growing wild from the

> rank productive force of the human mind, is almost tantamount, in the moral world, to the destruction of the apparently active properties of bodies in the material. It would be like the attempt to destroy (if it were in our competence to destroy) the expansive force of fixed air in nitre, or the power of steam, or of electricity, or of magnetism. These energies always existed in nature, and they were always discernible. They seemed, some of them unserviceable, some noxious, some no better than a sport to children; until contemplative ability, combining with practic skill, tamed their wild nature, subdued them to use, and rendered them at once the most powerful and the most tractable agents, in subservience to the great views and designs of men. (154)

Burke develops this line of argument to show that the methodical attempt by the new French regime at annihilating religious institutions in their country is an inept policy. To him, the Church is one of those "powers," of those "purchases" of the social and political world, which politicians should efficiently put to use: it may not only support but lend some energy to the whole machinery of the State and set it into motion. From this point of view, political scientists are the mechanical engineers of human societies: they should not start from scratch but humbly shape the world they are given.

Conclusion

Burke was a sort of prophet in the *Reflections:* he was right to warn scientists of the dangers of the Revolution. In 1793, when the Terror broke out, the most brilliant among them (Bailly, Condorcet, Lavoisier...) were condemned to the scaffold. Six years later, James Gillray indirectly illustrated this disavowal with his cartoon *Science in the Pillory* (6 March 1799)[14]. The scene evokes Napoleon's campaigns in Egypt, and the difficulties met by French scientists to perform their investigations among a hostile population. Like Burke, Gillray makes fun of the nonsensical French inventions. His work perpetuates, in the English conservative rhetoric, a scientific contempt for France, viewed as a country of Icaruses, of impractical scientists. The Montgolfiers' balloon has been blown up by Egyptian guns in Gillray's cartoon, and looks about to crash, echoing Burke's remark: "Let us be satisfied to admire, rather than attempt to follow, in their desperate flights, the aeronauts of France!" (244).

Bibliography

Barthélémy, Guy. 1988. *Les Savants sous la Révolution.* Le Mans: Cénomane,

Burke, Edmund. 1964. *Reflections on the Revolution in France.* London: Dent. [1790]

Chiron, Yves. 1987. *Edmund Burke et la Révolution française.* Paris: Téqui.

Col, Norbert. 2001. *Burke, le contrat social et les révolutions*. Rennes: PU Rennes.

Dhombres, Nicole. 1989. *Les Savants en Révolution, 1789-1799.* Calmann-Lévy: Cité des Sciences et de l'Industrie.

Fuchs, Michel. 1996. *Edmund Burke, Ireland and the Fashioning of the Self.* Oxford: Voltaire Foundation.

Ganzin, Michel. 1969. *La pensée politique d'Edmund Burke.* Thèse soutenue à l'Université d'Aix-Marseille, février.

Harris, R.W. 1963. *Political Ideas, 1760-1792.* London: Gollancz.

Shelley, Mary. 2012. *Frankenstein.* London: Penguin English Library. [1818]

Thierry, Patrick. 2010. *Burke, le futur en héritage*. Paris: Michalon.

Swift, Jonathan. 1973. *The Writings of Jonathan Swift.* New York: Norton.

Van de Vyver, Gisèle, et Jacques Reisse. 1991. *Les Savants et la politique à la fin du XVIIIe siècle.* Bruxelles: Editions de l'Université de Bruxelles.

Whale, John ed. 2000. *Edmund Burke's* Reflections on the Revolution in France: *New Interdisciplinary Essays*. Manchester: Manchester UP.

Notes

[1] In two famous speeches, *On American Taxation* (1774) and *On Moving His Resolutions for Conciliation with the Colonies* (1775), Burke had denounced the government's irresponsible policy in the American colonies. In 1784, he had tried to impeach Warren Hastings, governor of India, for his corrupted practice of colonial rule.

[2] "It was a great feast to me who had never heard him before. It was astonishing how all kinds of figure of speech crowded upon him. He was like a man in an orchard where boughs loaded with fruit hung around him, and he pulled apples as fast as he pleased and pelted the Ministry." (Peter Martin. *A Life of James Boswell.* Yale UP, 2002. p 294)

[3] The general contempt for the old science of alchemy in England in the last decade of the eighteenth century is well expressed in Mary Shelley's novel *Frankenstein* (1818) through the attitude of Victor's father who discards the works of Cornelius

Agrippa and Albertus Magnus as "sad trash" (31).

[4] The *Reflections* even contain a parody of Moliere's *Malade Imaginaire*: "Are the old *assignats* depreciated at market? —what is the remedy? Issue new *assignats— Mais si maladia, opiniatria, non vult se garire, quid illi facere? assignare—postea assignare; ensuita assignare.* The word is a trifle altered. The Latin of your present doctors may be better than that of your old comedy; their wisdom and the variety of their resources are the same." (231)

[5] "anodyne draught of oblivion, thus drugged, is well calculated to preserve a galling wakefulness, and to feed the living ulcer of a corroding memory" (68).

[6] James Gillray, *Copenhagen House,* 16 November 1795, http://www.npg.org .uk/collections/search/portrait/mw61929/Copenhagen-house, accessed Jan. 21, 2016.

[7] Gerrit Dou, *The Quack*, 1652, Museum Boijmans van Beunigen, Rotterdam, http://collectie.boijmans.nl/en/collection/st-4, accessed Jan. 21, 2016.

[8] This passage might be compared with several pages from *Utopia* (1516): with the geographical description given at the beginning of Book 2, but also with the presentation of the Utopian magistracy where we find a similar movement of successive subdivisions: 'Thirty families choose every year a magistrate, who was anciently called the syphogrant, but is now called the philarch; and over every ten syphogrants, with the families subject to them, there is another magistrate, who was anciently called the tranibor, but of late the archphilarch. All the syphogrants, who are in number 200, choose the Prince out of a list of four, who are named by the people of the four divisions of the city' (Utopia, book 2, 'Of their Magistrates').

[9] Alexander Pope's famous epitaph (1730) had turned Newton into a semi-god, while James Thomson's *Lines to the Memory of Sir Isaac Newton* (1727) had encouraged the circulation of the scientist's ideas.

[10] For instance when he compares the French revolutionaries to taxidermists who have disembowelled their fellow-citizens, emptying them of their hearts and entrails to serve their scientific aims of a political utopia, and concludes on a praise of British coarseness, and to some extent obscurantism : "In England we have not yet been completely embowelled of our natural entrails; we still feel within us [...] We have not been drawn and trussed, in order that we may be filled, like stuffed birds in a museum, with chaff and rags and paltry blurred shreds of paper about the rights of man. […] You see, Sir, that in this enlightened age I am bold enough to confess, that we are generally men of untaught feelings..." (83-84).

[11] Burke insists on the role of the Church in the "accumulation of vast libraries", and "collections of the specimens of nature, which […], by exciting curiosity, open the avenues to science" (58).

[12] James Gillray, *The zenith of French glory; -the pinnacle of liberty*, 1793, http://www.britishmuseum.org/research/collection_online/collection_object_details .aspx?objectId=1477710&partId=1, accessed Jan. 21, 2016.

[13] "And first of all, the science of jurisprudence, the pride of the human intellect, which, with all its defects, redundancies, and errors, is the collected reason of ages..." (92); "In history a great volume is unrolled for our instruction, drawing the

materials of future wisdom from the past errors and infirmities of mankind." (137)
[14] Gillray, *Siege de la Colonne de Pompée–Science in the pillory*, http://www.britishmuseum.org/research/collection_online/collection_object_details.aspx?objectId=1644150&partId=1&people=168164&peoA=168164-1-9&sortBy=fromDateDesc&page=1, accessed Jan. 21, 2016.

Doctor Jekyll and Citizen Stevenson: Figures of Science as Forms of Political Discourse in the Work of Robert Louis Stevenson

Cyril Besson

To this day, there persists a distrust between literature and politics. A political view of literature is seen as a trivialization of its formal dimension, making it subservient to social or political discourse; the literary view, in turn, is bound to convey, under the guise of formal preoccupations, the interests of a self-appointed social elite. According to John Lucas, confronting political issues "makes formal demands on the poetry and fiction in which it [this confrontation] strives for utterance" (1), although he advocates for a work-specific, rather than a systematic, approach. More evident perhaps is the trace of the political climate of the times in the literature of a certain era. The nineteenth century, especially, is a period whose politics is easier to approach as mediated through the fiction of its most eminent authors: the social commentary and call to reform offered by Dickens, the radicalism of George Eliot, and the changeable views of Robert Louis Stevenson.

Stevenson's political leanings fluctuated, from his early liberalism in Scotland to a form of anti-colonialism in the face of the imperialisms of the age in his Pacific days, although his Unionism and conservatism have often been too conveniently dismissed. As C. Harvie reminds us, as with most of his contemporaries, a subdued fear of an "inglorious" revolution haunted Stevenson, an author who was very much aware of the precarious position of the writer of standing: in his work, and in the nineteenth century in general, "the fear inspired by the growth of a socialist movement was mingled with a realization that artistic and literary culture had always been the province of a minority, and would inevitably be a casualty" (Harvie, 122). John Lucas further enters into the rationale of such apparently baseless fear: even though "it would be silly to pretend that there was no cause for fear in the early 1880s" (177),

> in 1880 there was practically no chance of organized revolution in England, and from this point of view the near-panic of politicians and social observers may seem absurd. But what we need to note is that the fears were capable of exaggeration because of a sudden recognition that all touch had once more been lost with the working-class and whatever forces were committed to its interests. Emblematic of this panic and ignorance is the lumping together of names like anarchist, radical, democrat and socialist, as though they all mean the same thing. (175)

Faced with the fearful prospect of revolution, for late Victorians, the mind boggled. This Stevenson epitomizes as he "muses […] on the emergence of a European anarchist movement. Perplexed by the anarchists' combination of 'dastardly' conduct and elevated 'spiritual life', he conjectures that '[t]his is just what the early Christians must have seemed to the Romans. Is it then a new *Drive* among the monkeys?'" (in Reid 175, italics original). In the "lumping together" of the Romans, the Christians, anarchists and a Darwinian echo within just one short sentence, one feels a sort of conceptual panic, indicative of a sense of the advent of a paradigm shift, but no less inconsistent for all that.

This, if it were necessary, establishes that Stevenson was of man of his time; of course, his life-time success is already proof enough that he represented something of the Victorian era, as attested by the emergence of the figure, "Stevenson." For it is not just the author who felicitously met with posterity, but also the public character, as Christopher Harvie highlights:

> The astonishing reputation of Robert Louis Stevenson, for a couple of decades after his death, were based as much on his life—and not simply his efforts to transcend ill-health, but on the ironies of his circumstances—as on his actual output. […] Stevenson was a phenomenon, and spawned the sort of literature we associate with phenomena, like science-fiction. […] Stevenson was distanced from the great Victorian tradition of high-minded social analysis. […] Kipling, Wilde, Shaw and Yeats, who shared some of Stevenson's 'phenomenon' quality, were in this respect within this tradition, Stevenson was an outsider. Like them he made his entry during a critical period in taste and organization, but his popularity remained a specifically literary one. (107)

It is not that Stevenson shied away from the controversies of the day, but his fiction, more than his essays and polemical writings, found a real echo with the Victorian public, though not so great as we imagine (Harvie 108).

Even qualified, the Scotsman's success was enough to make him a popular author, leaving him in a predicament: the uneasy negotiation between popularity and literary worth, two provinces seen as conflicting. Yet at the time, he was not as easily discarded as a writer of boys' adventure stories as he too often was at the end of the twentieth century. S*tyle* was retained of the author, Stevenson—including by himself. For Harvie, "by the 1870s, the enthusiasm of the more literary of the academic liberal group [Stevenson's patrons early in his career] for politics was dwindling fast" (110), and style, or at least the profession and general recognition of one's preference for it, appeared as a refuge. Here too, Stevenson is a man of his times:

> This disillusion with politics [of the 1870s] could manifest itself in conservatism or quietism, or in a growing professionalism. [...] *'Style', therefore, was not simply a literary affectation: it was a political declaration*—not far from 'art for art's sake'—a move away from a 'high seriousness' which found itself more and more perplexed by an intractable political situation. (110-1; italics mine)

Reputation was based on one's literary qualities, and one was accountable for it to a small coterie; with *popularity* came social responsibility, and perhaps the fear of being estranged from one's own work.[1] The mass of readers was construed as a polity of sorts, but certainly not one condoned by the *intelligentsia*. As Julia Reid writes:

> For Stevenson, popular literature is particularly dangerous, apt to release potentially contagious desires. Thus his valorization of the 'romance' is double edged, evoking its degenerate as much as its invigorating effects; it appeals to the lower instincts, gratifying the 'sensual' and 'illogical' tendencies of the human mind. The uncritical and sympathetic relationship to the narrative which Stevenson attributes to romance readers resonates with contemporary representations of the potential hazards of reading. It was a period when increasing emphasis was being laid on *how* to read, on the importance of thinking critically rather than reading addictively. Commentators construed the physiological involvement of the reader with the text as a possible danger. Discussions of hysteria and nervous diseases frequently cite the harmful effect of reading, demonstrating the widespread belief in 'the affective qualities of print'. [...] Critics viewed mass literacy and education, cheap literature, and a growing newspaper press as both causes and symptoms of degeneracy. (73-74)

Taken as a phenomenon, the boys' adventure story (the part of Stevenson's work most appealing to the buying public), entailed the promotion of "pro-social" values. Seen in this light, the specificity of Stevenson's work ran the risk of becoming indistinct. In texts like R. M. Ballantyne's *The Coral Island* (1858), moral discourse, not style, is perceived to matter, making literature subservient to the values being promoted, aligning the author with his society, at the expense of his literary identity:

> The nineteenth-century boys' adventure story had a didactic, conservative impulse […]. Its rewards were those offered by a stable society: financial security, if the hero survived, the conferment of moral worth, if he died bravely. The 'adventurer' and the romantic criminal—the highwayman, for example—with their satisfactions of momentary affluence and sex, were dislodged from a position they had earlier held in polite literature. (Harvie, 118)

For all his professions of being a writer of "romances," Stevenson took no small part in this change of nature in society's appreciation of the exceptional individual, but he introduced an ambivalent fascination that redeems the moralistic aspect of his adventure stories:

> Stevenson's criminals are not at all romantic […] although the ostensible heroes are worthy and resourceful in the manner of Marryat, Ballantyne and Henry. In this sense the stories of Jim Hawkins or David Balfour were appropriate pabulum for the middle-class child whom the educational system enjoined to be manly and plucky (frequently with practical ends in view), and even the appearance alongside them of more ambiguous adult figures —the attractive rogue (Silver) and the gallant with weaknesses (Alan Breck) were useful means of throwing the youth's necessary transition to 'independence' into relief. What, however, constitutes Stevenson's fascination is the welling-up of powerful figures from the subconscious which distort this simple didacticism, in much the same way that Jekyll's attempt to eliminate his animal impulses produced Hyde. […] it gains force when seen in terms of a social reality, which cut across and ultimately replaced the structure of the 'romance'. (Harvie, 119)

A correct understanding of the singular nature of Stevenson's work, therefore, rests on the appreciation of his ambivalence to popularity, and on the potency and complexity of the figures found in his fiction, which articulate social and cultural anxieties in an idiosyncratic manner.

Among those figures are not just "exceptional individuals"; a good deal of them are much closer to the common man, or at least the common Victorian, than Long John Silver or Henry Durie. Undoubtedly, a number of the most remarkable figures in Stevenson's work are anchored in the social reality of professionalism: from Utterson to Rankeillor, lawyers abound in Stevenson's work, and their overbearing presence does indicate his obsession with the theme of law and the social responsibility of the individual to society. Of course, the most striking professional figure, that which truly captured the public imagination, is none other than a certain doctor, a character to rank with Frankenstein and Moreau in the pantheon of the so-called "mad" scientists of literature. Though eclipsed by Henry Jekyll and his alter ego Edward Hyde, there are, however, other practitioners of science in the Scotsman's texts: doctors, to be sure, but also engineers, perhaps not scientists per se but figures of science. If Stevenson's uncompleted training as engineer was put to little use, and he cannot really be said to have significantly entered the scientific controversies of his day, as his preoccupations were far more socio-political, science *did* find its way into his miscellaneous writings, as theme and backdrop, and was used (and perhaps, indeed, abused) to express something of the spirit of the times.

It will be the contention of this paper that Stevenson figures, and figuratively appropriates for himself, not just positions available to practitioners of science in society, but a whole array of "figures of science," from the most concrete (the physician, an occupation of particular import for someone of such constant poor health as Stevenson), to the most abstract, so abstract in fact that "pure science" as a speculative force is never fully embodied into a character or professional in his work, but is present as pure *drive*. Doctors and engineers, as social functions more than as practitioners of science, materialize into the pages of Stevenson's writings in ways that the physicist or anthropologist cannot. Constituted as two socially visible and effective corps, they were manageable figures for Stevenson; they had already been appropriated by culture, be it general society (the many doctors of fiction) or family myth (his own forebears, a family of engineers, "the Lighthouse Stevensons"). The pre-existence of these models made them relatively easy to reconfigure and appropriate, and they came to embody for him notions of authority–almost systematically deficient for doctors, more assertive and even commandeering for engineers. They are in this sense, figures of power, presumably symbolic of the views of his regular readers, expressive of a social and political function that delineates a certain vision of the privileges and responsibilities of the professional middle class.

This of course is not specific to Stevenson. Scientists are often retained as cultural artifacts, even outside the self-conscious exercise of fiction. As exemplified by Stevenson's biographies of his grandfather and his friend and former teacher, Fleeming Jenkin, accounts of the lives of scientists tend to be more interested in the exceptional character of the man than in the specificities of his work. Some scientists were complicit in that social mystique of "genius." As Tess Cosslett writes: "The language of genius was also taken up by those in scientific circles more strictly defined, the pre-eminent example being Humphry Davy. Certain biographical traits were commonly picked up in accounts of students in nature" especially "the ability to stick with ideas, even in the face of opposition, displaying bravery, tenaciousness, even a zeal, a passionate commitment" (70).

These accounts of students of nature were the chief means through which science circulated in the social imagination, but their idealization of scientists align them with fiction more than with true biographies. In the words of Chassay, "social reality tends to make myths of their [the scientists'] life and work" (17).[2] As he says, "Science 'performs' science, scientific vulgarization tries to explain it, and fiction stages it" (17) allowing the historian to offer "a reading of those contemporary crises produced by or because of the sciences [that] allows one to ponder the structures of those modes of thinking that are still ours […]. Fiction does not replace reality, it only intensifies it" (17). As mediated by culture, "certain great characters, owing to the stance that they took or the scope of the debates that their findings triggered, polarize the *cultural effects* of science. Fiction often insists on the ravings, fantasies, irrational beliefs or the infinitely puzzling musings they cause" (18, italics original) and in so doing, "it helps us understand how science shapes or realizes itself, showing *how sciences think*, how they render our world and bear upon the way we think about it" (18; italics original). Scientific crises are always crises of representation, testifying to a shift in the vision of the order of the universe, and therefore, the rationale behind structures of power; the central figure of the scientist metonymically becomes a leader of men, though not necessarily on the path righteous.

As this essay shall establish, this is re-enacted in Stevenson's work, where in a first movement, some doctors *embody* the destructive potential of science, so disruptive of the world order that it might make it impossible to reconstruct; as a second movement, in his biographical writings, engineers implicitly rearrange the world, improving on the original, reinforcing order. Eventually, these two visions of science find themselves reconciled in the conception that emerges from Stevenson's most abstract writing on science, the short essay "Pulvis et Umbra," providing the reader with an unconventional stance on the way science should be wielded for the good of mankind.

First, do not harm? Doctors in Stevenson's Fictional Work

In Stevenson's fictional writings, doctors are inept therapists. Livesey from *Treasure Island* matters more as a lawyer than as a physician, a job he is only competent at; nobody is cured in "The Body Snatcher," his tale inspired by the Robert Knox / Burke and Hare case (1884); Jekyll's practice is notoriously limited to a single patient, himself, with devastating results. His exercise of medicine is not really science, but alchemy, and even then, his partial success is only due to the defect of one of the chemicals he uses. Jekyll is the best of these unheroic figures whose professionalism is never made much of by the texts. The contrast with lawyers is striking, as *they* are shown to master their "art," especially during battles of wits where their science of their trade is on full display. There, perhaps, lies a surprise for the reader well-versed in the professional practices of the Victorian era: what solitary figures Stevenson's doctors are! They are never quite presented as a consistent "body"—like Jekyll in the text, whose "full statement" is no less on the fringe of the novella than *he* eventually is on the fringe of his own body, peripheral and isolated.

This isolation, with the pessimism that comes with science in the Scotsman's work, stands in opposition to the path of the medical profession in the nineteenth century. By the publication in 1818 of Mary Shelley's *Frankenstein*, an indirect progenitor of *The Strange Case...*, this image was already a thing of the past, as Crosbie Smith reminds us: Frankenstein is "an 'extraordinary' rather than conventional man of science" and "Frankenstein's obsession, his isolation, his individualism and his egoism are strongly suggestive of Romantic images of the mad genius, the creative artist, and the natural philosopher *qua* natural magician" (41). For Ludmilla Jordanova, provincial medical societies had from the 1770s undertaken a professionalization of the vocation, working "against the grain of most medical practice, which was solitary, and carried out in domestic settings" (63).

By 1886 and *The Strange Case...*, the solitary doctor, left to his own devices, unsupported or unsupervised by the other members of his corps, was a thing of the past, so that Jekyll's seclusion is not a historically accurate notion. In fact, Jekyll is not left unchecked in the text, but is confronted to Dr. Lanyon, his foil and nemesis, overlooked by many a reader. He is the only one to hear Jekyll's confession, and is so smitten with the vision of the transformation into Hyde that he dies from the shock retrospectively. A true professional who finds Jekyll's ideas "too fanciful", a materialist and rationalist in the face of Jekyll's quasi-mystical quest for undiluted moral purity, Lanyon embodies, however ineffectively, the normative reaction of middle class professionals as a repressive mechanism of control against the potential overindulgence of some practitioners. He is a reminder that doctors formed a body in Victorian society, and that the individual practitioner was not allowed to exercise his trade unregimented. The individual physician was duty-bound to represent his peers, and L. Jordanova aptly summarizes this mutually beneficial interplay between the individual and the "institution," the latter presenting "the public face of science/medicine as a collective enterprise" so that "the making of individuals into scientific heroes and […] institutionalization" are "complementary faces of the same coin" (64).

This intraprofessional self-consciousness characterized practitioners of medicine in their own eyes and in those of society, and gave them authority, social power in Bourdieu's sense. At a time when the notion of "the scientist" had not taken form in any *finalized* sense, doctors were the forerunners of this elusive concept, "the scientific community." They were "science heroes", a secularized version of the over-archers that followed in the footsteps of Faust, those ambiguous antiheroes bound to be censured in the Christian mindset, except now rendered morally acceptable by society because they "did no harm." To quote Jordanova again, if "[i]t was important that the power flowing from natural knowledge was purged of its magical and hubristic elements" (72),

> it was attractive to create a certain aura around scientific and medical activities that presented them not just as worthwhile in terms of contributing to the progress of knowledge and human well-being, but as thrilling. A good deal of mythologizing was involved in terms already established as plausible at this particular historical moment. None the less, this image/self-image was fragile, it was delicately poised between social benefits deriving from knowledge well used, and disasters derived from certain kinds of excess. Practitioners had recourse to a range of devices to cultivate the former and keep at bay the latter. Histories of their subject did just this. These were quite explicitly about fatherhood in both its good and bad forms. […] Other devices included the elaboration of intraprofessional etiquette, cultivation of patronage relations, and assertions of moral and/or religious conformity." (72)

We will touch upon the histories of the scientists' subject in further detail when we tackle Stevenson's biographical accounts of the lives of various engineers, but let us for the moment emphasize how true it is that filiation, although not always literal, is of the essence for the medical profession. It certainly is for Hyde, confronted to an impossible choice between the creature that he, a father and mother figure in a process of hermaphroditic parturition, carries within him, and "Utter-son," the typical male progeny in Stevenson's fiction, whose obsession is to kill the father and be left as sole inheritor, in this (legal) *case* by appropriating Jekyll's capital for himself. Devoid of a professional ethos, as established by the fact that in his (medical) *case*, even the most basic operative rule of the medical profession, the separation between doctor and patient, is not respected, Jekyll therefore becomes the locus of all confusions. With him, all established notions are displaced, and the figure of the doctor, at the time well on its way to becoming stabilized in social representation, devolves into an untenable Gordian knot of anxieties that stands against the realities of the medical profession, paradoxically attesting to what they were: everything that Jekyll was not.

Repudiating intraprofessional solidarity, free from any network of medical associations, he is not what any professional is expected to be: a repository of the cumulative knowledge of his trade. Neither does he contribute to the building-up of the knowledge of his peers, as he creates the conditions for his scientific legacy *not* to survive him. The secrecy of his laboratory is absolute; it is not put to good use in scientific terms, if we consider what a laboratory is for, not in the public imagination, but for the scientific community. It is not an ivory tower abstracted from the world, even if it is true that it is an artificial environment isolating the scientist in the day-to-day exercise of his vocation. Though not exactly a holy of holies, it still requires the exclusion of the profane, and bans contamination from the external world, testifying to the profession's peculiar connection to the rest of the social field. Thomas Kuhn sums this up quite neatly, exposing "the unparalleled insulation of mature scientific communities from the demands of the laity and of everyday life. […] [T]here are no other professional communities in which individual creative work is so exclusively addressed to and evaluated by other members of the profession" (163). Of course, this summarizes a situation that was only developing in Stevenson's time, but it does entail that practitioners of science see themselves as *by nature* part of a network. This "web" is certainly more diffuse and immaterial than the ivory tower vision of the laboratory in the popular imagination. Outside the lab, a "republic of the sciences," a polity of sorts, has been in existence for at least since Newton,[3] through the publicity of exchanges in forums of different sorts. Politically speaking, this polity's constitutive exclusiveness makes it not a democracy but a meritocracy; Jekyll's isolation only radicalizes the fact. The flipside of the coin is that the common man does not have access to the intricacies and sophistication of scientific reasoning, and is exposed to scientific theory only insofar as it induces changes in the worldview pervasive in his or her society —heliocentrism is a case in point, as Copernicus's calculations mattered much less to his contemporaries than their theological consequences.

What matters to the bulk of society, when it comes to science, is therefore not the heuristics, but the epistemology. Society reappropriates science for itself, translating it into a symbolic currency it can circulate within its bounds, in its order. Scientific crises are crises of representation, metonymically indicating shifts in the way the order of the "universe" is envisaged, and therefore, the motivation behind structures of power— social Darwinism is a case in point, regardless of what Darwin himself had to say. Willy-nilly, the innovative scientist is made into a standard bearer by those members of society trying to do away with traditional modes or agents of power, leading to changes that are political in nature. Politics takes advantage of science, using it improperly, by analogy, to justify itself, especially when new ground is being broken. These two orders, science and politics, are in fact competitors for supremacy in the social field.

This opposition is dramatized in *The Strange Case*.... The murder of Danvers Carew, a Member of Parliament, is an actualization of this conflict of "professionals" belonging to different fields, each one thinking itself the more useful and "organic" to society. For S. Arata (1995, 233), the middle-class professionals "close ranks" from the very beginning of the novel, when Utterson and others of his class protect Hyde against the vengeful fury of women made mad by his trampling of a little girl. Utterson and co. symbolically accept Hyde as one of their own, and he enacts a number of unfulfilled frustrations on behalf of this class. Murder becomes a social syndrome, indicative of the conflict for pre-eminence between the political and scientific fields. Of course, Jekyll is in a way infra-political, so obsessed is he with himself and himself only, even in another avatar; he is also beneath the "normal scientist"[4], not a member of the scientific *community*—in a class all his own. He nonetheless brings to the fore, in reverse, all the norms of the medical community, and allows Stevenson to examine, against the grain, the strategies that presided over its constitution. For A. Stiles:

> By making his diseased protagonist a physician, Stevenson ironizes the power relations of nineteenth-century medicine. […] his focus on the "strange case" of a male professional implicitly attacks the prerogatives of a patriarchal institution that for the most part […] treat female subjects. […] In a century that constructed madness as stereotypically feminine, male doctors appeared not only as the possessors, but also as the dispensers, of reason, which they can at will mete out to—or take away from—others. Stevenson's male doctors struggling to repress the feminine elements within himself destabilizes the paradigmatic power relations, between male physician and female patient, as well as the identification of women with madness that Showalter so thoroughly elucidates. (895)

This is confirmed by Julia Reid: "While British psychologists still diagnosed hysteria as an almost exclusively female disorder, there was a growing tendency to pathologize cases of male nervous susceptibility, and to brand the nervous male effeminate" (63). Jekyll, characterized by anxiety, comes to embody the pathological case of "male nervous susceptibility," and his "androgynous" status as both father and mother to an unhealthy child that is not quite "not-himself" challenges the social basis for the authority of the man of science, the credibility granted to the man in control of himself.

This, then, is the political work undertaken by Stevenson in *The Strange Case...*, his political stance on science: shedding light onto relations of power within society, and exposing the figure of the doctor, as illustrative of that of the scientist, and its demands on the individual. This discursive "politicization" of the figure allows us to use the categories of political discourse in Aristotelian thought[5] to examine the shift that the figure of the scientist undergoes under Stevenson's pen. By presenting, in *The Strange Case...* especially, but also in "The Body Snatcher," the dereliction of intraprofessional solidarity in the medical field, Stevenson first denaturalizes the modes of authority constitution (*ethos*) in the field itself. By choosing to focus on the shillyshallying of his medical practitioners (doctors or the peculiar "orderlies" from "The Body Snatcher"), he transposes them as creatures not of *logos*, but of *pathos*, bringing about emotion in the receiver, reconfiguring our perception of the figure of the medical scientist. Medical practitioners in Stevenson's fiction achieve a reshaping of the parameters inherent in scientific discourse, as exemplified by the attribution to Jekyll of virtually no *logos*, the reexamination of his *ethos* and the extolling of his *pathos*, resulting in a new figure, and possibly a new role in society, for the scientist.

"The romance of engineering"

This reexamination of the figure of the scientist through doctors contrasts sharply with the ways in which Stevenson deals with another major figure, the engineer, conspicuously absent from his *fictional* writings. Coming from a long line of engineers, he may have found it too "close to home," and his somewhat adversarial relationship with his father precluded him to launch a criticism of the same nature as that which he undertook on the figure of the doctor.

Engineering, the profession of the family line, stood for the idealized father for him, exemplifying a conservative patriarchy that imaginatively guaranteed a place for Stevenson with a corps of professionals, especially in the face of the criticisms levelled at his claims to the status of professional *littérateur* when he established himself in Samoa. Doctors were different, creatures enabling him to appropriate science for himself in the context of fiction. He did with them what he could not do with engineers: apply their transformative abilities to them. Jekyll especially, a "plastic" being, is a material which does not resist the application of science as regulated by the author. In Allen MacDuffie's reading, "*Dr. Jekyll and Mr. Hyde* is a novel acutely concerned with the transformation of physical forms, the scarcity of resources, and the directionality of the universe. Jekyll's is, at heart, an experiment in human engineering, one that unlocks Hyde's 'energy of life'" (3). The novella is a work that draws "upon the principles of engineering to direct not only the course of the physical metamorphosis but also the shape of the narrative. The irreversibility of physical transformations represents one of the most critical points of disjunction between "classical" and modern science, and it informs Stevenson's modern rewriting of Ovidean metamorphic tropes and structures of allegorical representation" (3). *Jekyll and Hyde*, then, "is a metamorphosis updated for the industrial age, for a world in which great transformative capacity was at the command of engineers, industrialists, and, increasingly, ordinary people" (3). Hyde is the end result of his society, *embodying* a scientific and social revolution, while the engineers in the end have a very superficial relation to science, using it as symbolic capital to assert their authority.

However, the engineer is no less a repository of ambiguities in Stevenson's biographical writings, discreetly enabling him to contrast himself and his various endeavours with the achievements of his forefathers. Thus, in *Random memories,*[6] the narration of his personal experience of engineering implicitly tells the failure of an education in the patrilineal profession. The last and most revealing part of *Records of a Family of Engineers*,[7] Stevenson's attempt at a family history, is his grandfather's diary, framed and censored in striking ways. Robert Louis's eulogy of his father ("Thomas Stevenson, Civil Engineer") is another case in point.

In all these, the at times dithyrambic presentation of the engineer as a creature of civic duty, the embodiment of benevolent *paternalistic* authority, entirely devoted to the good of his society in unclear ways, obscures the fact that the "province of the engineer" (Stevenson's phrase) is subtly bounded by those texts. This genus, also a genius,[8] belongs to a certain order, in the many senses of that word, and cannot assume political power in just about every sphere, for fear of losing its specificity. The engineer's authority rests on his expertise, the form of power he can exert in "his" world is absolute; beyond that realm, his legitimacy is questionable. Other figures of the expert are valid in their own order, most visibly the true *littérateur*; a *pro domo* argument, no doubt. This, however, was Stevenson's way of resisting the increasingly technocratic mode of power of society, "[s]o that ultimately Stevenson's political creed is authoritarian but–unlike Kipling–feudal and familial rather than technocratic" (Harvie, 124).

The story of Stevenson's first foray into engineering appears as an initiation into a corps that comes to a sudden end–an abortive *Bildungsroman*. When it is decided that the little boy of an oversensitive disposition should be made to "man up" by following, almost literally, in his father's footsteps (*Random Memories,* 112), the required absence of "the petticoats" presents a relationship to femininity and fear of masculine deficiency in the context of patriarchy quite similar to *The Strange Case…*, though certainly mediated by Stevenson's constant irony in the essay. But this surrogate education, into manhood this time, is presented as ineffective: "Hanging about with the east wind humming in my teeth, and my hands (I make no doubt) in my pockets, I looked for the first time upon that tragi-comedy of the visiting engineer which I have seen so often re-enacted on a more important stage" (119).

The apparent deflation of the figure is mitigated by the fact that, tragi-comic or not, the engineer's visit to a lighthouse is a requirement, an imperious necessity completely superseding the individual wills involved. An obligation due to the very dangerous consequences of negligence, the engineer's exercise of his expertise is shown as other than the practice of applied science, creating a situation where power is exerted, in the name of maintenance–in symbolical terms, the preservation of order.

> Eighty years ago, I find my grandfather writing: "It is the most painful thing that can occur to me to have a correspondence of this kind with any of the keepers, and when I come to the Light House, instead of having the satisfaction to meet them with approbation and welcome their Family, it is distressing when one is obliged to put on a most angry countenance and demeanour." This painful obligation has been hereditary in my race. I have myself, on a perfectly amateur and unauthorised inspection of Turnberry Point, bent my brows upon the keeper on the question of storm-panes; and felt a keen pang of self-reproach, when we went down stairs again and I found he was making a coffin for his infant child; and then regained my equanimity with the thought that I had done the man a service, and when the proper inspector came, he would be readier with his panes. The human race is perhaps credited with more duplicity than it deserves. The visitation of a lighthouse at least is a business of the most transparent nature. As soon as the boat grates on the shore, and the keepers step forward in their

> uniformed coats, the very slouch of the fellows' shoulders tells their story, and the engineer may begin at once to assume his "angry countenance." Certainly the brass of the handrail will be clouded; and if the brass be not immaculate, certainly all will be to match [...]. If a light is not rather more than middling good, it will be radically bad. Mediocrity (except in literature) appears to be unattainable by man. (117)

No longer just a technical specialist, the engineer becomes an expert in men, "seeing through" them. In keeping with the Calvinist tradition, sloppiness becomes an index of a man's character, and the engineer himself, a moralist at a glance. For all his reticence, the modality of the engineer's punitive responsibility is deontic; he can never be wrong in the exercise of it, even when he is inhumane. *Dura lex sed lex*–such is the engineer's creed, and the gentleman's burden. There is something of the aristocrat about this figure of the man exercising his authority for the good of the Body Politic because it runs in his veins. The atavistic necessity of power relations in this context is not set into question, and this figure does not undergo the examination that the scientist as doctor is subjected to. Contrary to Jekyll's case, the absence of a professional network in those works of Stevenson's that deal with engineers (and not really engineering) is not presented as a problem; no trace of intraprofessional solidarity is required, for the engineer is a political figure sufficient onto himself. The reality, of course, is very different. According to Yves Gingras (113), the engineer finds himself at the centre of the nexus of the "technological system," but although essential, he is just a link in the chain. Never is this exposed in the Scotsman's work, which prefers to offer a mystique of the engineer, presented, if not as the originator of all things, at least as God-like, oiling the wheels of Creation, after developing a skill particular to his profession. In Stevenson's memoir of his grandfather, measuring, weighing, numbering may be

> the certainties of the engineer; so far he finds a solid footing and clear views. But the province of formulas and constants is restricted. Even the mechanical engineer comes at last to an end of his figures, and must stand up, a practical man, face to face with the discrepancies of nature and the hiatuses of theory. After the machine is finished, and the steam turned on, the next is to drive it; and experience and an exquisite sympathy must teach him where a weight should be applied or a nut loosened. With the civil engineer, more properly so called [...], the obligation starts with the beginning. He is always the practical man. The rains, the winds and the waves, the complexity and the fitfulness of nature, are always before him. He has to deal with the unpredictable, with those forces [...] that 'are subject to no calculation'; and still he must predict, still calculate them, at

> his peril. His work is not yet in being, and he must foresee its influence: how it shall deflect the tide, exaggerate the waves, dam back the rain-water, or attract the thunderbolt. […] he must not only consider that which is, but that which may be.

Not one word on the social use of the machine or of the construction; they exist in a quasi-vacuum, not operating by themselves, but rather being operated, even imaginatively and prospectively at their inception, by the engineer only. What counts is the "running" of the thing produced by applied science, which, presented here as preserved from external interference, human and otherwise, seems to have the potential to realize that age-old fantasy of physics, perpetual motion (if the engineer designs it well enough, that is). The engineer is an enlightened being, accounting for nature as it is, not as we would have it, in a sense embracing all Creation. So enlightened is he that he anticipates by a century Nancy Cartwright's verdict that grand systems in physics only ever work *approximately*, and that corrections have to be brought to them constantly if one expects to make use of them:

> the engineer has need of some transcendental sense. […] The rules must be everywhere indeed; but they must everywhere be modified by this transcendental coefficient, everywhere bent to the impression of the trained eye and the *feelings* of the engineer. A sentiment of physical laws and of the scale of nature, which shall have been strong in the beginning and progressively fortified by observation, must be his guide in the last recourse.

A professional superman of multifarious acuity and acute sensibility ("sympathy"), the engineer is a quasi-religious figure, for the nature of his science is ineffable:

> But it is of the essence of this knowledge, or this knack of mind, to be largely incommunicable. 'It cannot be imparted to another,' says my father. The verbal casting-net is thrown in vain over these evanescent, inferential relations. Hence the insignificance of much engineering literature. So far as the science can be reduced to formulas or diagrams, the book is to the point; so far as the art depends on intimate study of the ways of nature, the author's words will too often be found vapid.

Engineering is not just a vocation, but a way of seeing the world; the secret of his art is for the happy few. He is a civil hero for the modern age of meritocracy, no less abstracted from his society than Jekyll, except his virtues make up for the potentially antisocial tendencies of this perceived isolation.[9] This means turning a blind eye to something conspicuously missing wherever the Scotsman tackles the subject: the building or inspections of lighthouses were *commissioned.* Instead of presenting the engineer as underling, Stevenson focuses on the exercise of authority "in the line of duty," or the qualities required of this figure—self-sacrifice especially. Nowhere is this more apparent than in Stevenson's "elegy" for his father, "Thomas Stevenson, Civil Engineer" (in *Memories and Portraits,* 129-139). He is presented as displaying an abnegation

> that applies to the whole family, and only in particular to Thomas Stevenson from the great number and importance of his inventions: holding as the Stevensons did a Government appointment, they regarded their original work as something due already to the nation, and none of them has ever taken a patent.[10] It is another cause of the comparative obscurity of the name: for a patent not only brings money, it infallibly spreads reputation; and my father's instruments enter anonymously into a hundred light-rooms, and are passed anonymously over in a hundred reports, where the least considerable patent would stand out and tell its author's story. (134)

In what reads like a panegyric, the figure that emerges is not the unknown soldier, but the unknown genius, a secularized saint, though no more a prophet in his own land than his religious counterpart: for all his achievements,

> to the general public and the world of London, except about the parliamentary committee-rooms, he remained unknown. All the time, his lights were in every part of the world, guiding the mariner; his firm were consulting engineers to the Indian, the New Zealand, and the Japanese Lighthouse boards, so that Edinburgh was a world centre for that branch of applied science […] ; even in France, where his claims were long denied, he was at last, on the occasion of the late Exposition, recognized and medalled. (130)

"[T]he inverted nature of his reputation, comparatively small at home, yet filling the world" (130), is the source of bitter regret for his son in this eulogy, and tellingly, in the next lines, he recounts an anecdote testifying to the fact that in some places, the name, "Stevenson," is bound to be patronymic in the etymological sense—associated to Thomas, not Robert Louis. After Thomas's death, their comparative fames cease to be a potential source of intergenerational competition,[11] and filial piety takes over.

Unlike Dr. Jekyll, an anti-hero destabilizing the figure of the scientist, the engineer in Stevenson's biographical writings is the embodiment of the unquestioned values of his class, and the authority of his moral character is taken for granted. There is no discrepancy between his public figure and his true self *in operative terms*, and he personifies the pragmatism (common-sense?) of applied science, no longer confronted to the parasitic resistance of the individual organism, but to the inertness of materials which are also of the human sort. Lighthouse keepers or workers are not shown to rebel against the engineer's authority, and there looms over them the suspicion that they are destined, if not to inertness, at least to inertia; their lack of care could be assimilated to a form of entropy, and just as in that state, it is necessary to re-inject energy from without the system (the social engine?) to restore it. By exercising his authority, the engineer is the operator of this socially salutary "inoculation."

A model of self-restraint, the engineer masks inner conflict, as Stevenson himself did when he realized on his inspection that the keeper's son had died. By abandoning the mask of the engineer (and it was one, as he was no engineer), he would have displayed the moral conflict in himself between the professional and the human being, undermining the appearance of assuredness and his authority. But he did not, and so there was no discrepancy between his public figure and his true self *in operative terms*. Now, as exemplified by countless novels, *The Strange Case...* among them, inner conflict is the core of fiction, and this equation of function and self in the case of the "engineer" type of scientist may explain why this figure is a subject Stevenson tackles only in his biographical writing, which at times reads like hagiography. Others have postulated different reasons for the relatively minor place of the engineer in fiction. Charles P. Snow, for one, surmises that perhaps the engineer is not a good subject for Western culture in general because "an engineer in a soviet novel is as acceptable, so it seems, as a psychiatrist in an American one" (Snow, 37), which presumably testifies to the different social priorities of the two cultures (social efficiency for one, the individual psyche for the other). Maybe so; however, if indeed it is an improper novelistic character in the West, as his absence from Stevenson's fiction illustrates, the figure of the engineer, with distinctive *ethos*, *logos*, and *pathos*, and a form of authority natural to his class, is not without a place in the cultural construction of science and political discourse.

"Of the Kosmos in the Last Resort, Science Reports Many Doubtful Things and All of Them Appalling"

In Stevenson and others' work, fiction and the less rigorous sort of biographies "present scientists whose features are hardly recognizable and whose activities are fictionalized, or even contaminated by the fantastic. They can be made to cross over into the realm of fantasy because of the logic of discovery. Unravelling the mysteries of the universe means placing oneself at God's level, which is of major import imaginatively speaking" (Chassay, 17). Of Stevenson's creations, perhaps only Hyde truly displays such hubristic and metaphysical ambition, but there is a pervasive sense that science never leaves the world intact, entailing a reconfiguration of man's position within it. Figures of science are transformative of reality, the greatest power there is in discourse. However, even in the case of Hyde's alchemical, rather than chemical, experiments, Stevenson's "handling of the supernatural is a conjuring trick: there is no sense that rational causality could actually be overthrown in the real world" (Harvie, 119). This unshakeable belief in causality, even one not fully controlled, is important, and aligns Stevenson with his contemporaries in terms of what worldview paradigm shifts entailed: the challenge to tradition was not proof of a universe ruled by chaos or *anarchy* of the natural sort, but rather, of a concealed order to be unveiled by those enlightened enough first to see it, and second, to accept it. As Tess Cosslett writes:

> The ideal of natural causation implies not only dependable regularity but also coherent *unity* in Nature. The tendency of Victorian science was to stress this oneness, by breaking down the barriers between organic and inorganic; Nature is unified by the universal operation of scientific laws; and also by the uniform 'material' composition. This uniformity can be presented as a depressing and ugly triumph of 'mechanism' and 'materialism', driving all 'poetic' beauty from Nature, and finally entrapping man himself in this alien machine. However, this is not the only interpretation. Huxley and Tyndall always insist that the new discoveries do not degrade organic life, but rather dignify so-called 'inorganic' matter. Tyndall, in particular, as a physicist, is continually aware of the energies and structural potentialities of 'matter', revealed by discoveries in thermodynamics, for example. So instead of the material analogy being extended upwards, the analogy of life could equally well be extended downwards, and the whole of Nature, including man, be seen as one living organism, rather than one dead machine. Instead of feeling an alien in a hostile universe, man can just as well have a new feeling of kinship with the rest of Nature. (17-18, italics original)

Beyond the mere facts, the worldview that emerges from one's contemplation of scientific results depends on subjectivity,[12] and it can go either way.[13] Cossett gives an example of this framing of science in "the main values attached to the theory of evolution": not atheistic nihilism as feared by its opponents, but rather "the acceptance of unpleasant Truth and renunciation of 'supernatural' consolations; the vision of unifying, all-pervasive Law; the recognition of kinship with Nature; the appreciation of interrelatedness and co-operation within Nature; and the admiration for the scientific imagination" (13).

The confrontation between the optimistic and pessimistic views resulting from scientific "revelation" shows through in the work of Stevenson, reintroducing looseness to science where society tends to fossilize it as indisputable truth. Stevenson contests this view spectacularly in the 1888 essay, "Pulvis et Umbra" (in *Across the Plains*, 193-201), the epitome of the "abstract" strand of the figuration of science in his writings. The first part of this "Darwinian sermon," as he called it, expresses in the impersonal mode the despair induced by science, or perhaps one should say, *omni*-science: through science, the piece's thinker (for want of a better word, but this discursive indeterminacy goes to show that no "professional" role is set for him—what *ethos* for *him*?) knows all, but all-pervasive powers of observation mar the very experience of the world.

> Of the Kosmos in the last resort, science reports many doubtful things and all of them appalling. There seems no substance to this solid globe on which we stamp: nothing but symbols and ratios. Symbols and ratios carry us and bring us forth and beat us down; gravity that swings the incommensurable suns and worlds through space, is but a figment varying inversely as the squares of distances; and the suns and worlds themselves, imponderable figures of abstraction, NH3 and H2O. Consideration dares not dwell upon this view; that way madness lies; science carries us into zones of speculation, where there is no habitable city for the mind of man. (193)

The universe is entirely knowable, but no more meaningful for all that. Man may measure all things, but is no longer himself the measure of all things; he and the universe are not commensurable, and the latter offers only cold data, deprived of any sense of order, harmony, or purpose. A few lines later, in the same vein, scientific reductionism is a source of despair: the stuff of all things is matter, a category that resists understanding, especially in its animated variety.

> All of these [the planets and stars] we take to be made of something we call matter: a thing which no analysis can help us to conceive; to whose incredible properties no familiarity can reconcile our minds. This stuff, when not purified by the lustration of fire, rots uncleanly into something we call life; seized through all its atoms with a pediculous malady; swelling in tumours that become independent, sometimes even (by an abhorrent prodigy) locomotory; one splitting into millions, millions cohering into one, as the malady proceeds through varying stages. This vital putrescence of the dust, used as we are to it, yet strikes us with occasional disgust, and the profusion of worms in a piece of ancient turf, or the air of a marsh darkened with insects, will sometimes check our breathing so that we aspire for cleaner places. But none is clean: the moving sand is infected with lice; the pure spring, where it bursts out of the mountain, is a mere issue of worms; even in the hard rock the crystal is forming. (ibid)

Nihilism looms large in these lines. "Evolution," although not yet of the Darwinian type (it will appear shortly), because it does not march steadfastly towards an end, is the cause of the omniscient "narrator's" disgust. All-knowing, he is bound by the impotence, not so much of his own flesh (he is pure spirit, although revulsion at the organic is given free rein in this piece), but rather of his mind to reconcile the totality of unstable phenomena with a teleological culmination. Why such toils, then, when destruction or fossilization would be preferable to a rational mind? *Quia absurdum*, Stevenson seems to answer halfway into the essay, and the specter of the pessimism induced by a perception of science as intrinsically unsatisfying is conjured away by a sleight of hand: the morality of man, Job-like in a godless world, unexpectedly finds itself rehabilitated. His duty is to himself and his fellow men—or creatures, or elements, even:

> A new doctrine [Darwinism], received with screams a little while ago by canting moralists, and still not properly worked into the body of our thoughts, lights us a step farther into the heart of this rough but noble universe. For nowadays the pride of man denies in vain his kinship with the original dust. He stands no longer like a thing apart. Close at his heels we see the dog, prince of another genius: and in him too, we see dumbly testified the same cultus of an unattainable ideal, the same constancy in failure. Does it stop with the dog? […] The whole creation groaneth and travaileth together. It is the common and the god-like law of life. The browsers, the biters, the barkers, the hairy coats of field and forest […] share with us the love of an ideal: strive like us—like us are tempted to grow weary of the struggle—to do well […]. (201)

This Sisyphus-like mentality is a paradoxical, secular, and indeed Darwinian leap of faith, a triumph of the will to overcome the deleterious effects of a science that threatens to destabilize not just traditional order, but human life itself. Not science, but what it entails for the human mind and for society, is what this, and much of Stevenson's discussion of science, is about. The effects of scientific progress are not universally good, and society should not accept them without thinking, as the essay "A Plea for Gas Lamps" (1878) demonstrates. There, the "new" energy of electricity, at least as applied to public lighting, is envisaged as poisonous for the soul, probably because it is operated by a "sedate electrician," and it does not entail the same mechanisms of control and sanction as those in place in the context of the lighthouse. Stevenson therefore predicates the use of science on a form of "political" control (in the broad sense of the word, although he does apply the epithet "conservative" to himself), and even on a social and moral aesthetics: "a new sort of urban star now shines out nightly, horrible, unearthly, obnoxious to the human eye; a light fit for a nightmare! Such a light as this should shine only on murders and public crime, or along the corridors of lunatic asylums, a horror to heighten horror" (277). In this case, progress seems uncontainable ("Such is the spectacle of the future…," 276), but the certainty should be mitigated with dithering at the costs for the integrity of society.

Conclusion

In Stevenson's fiction, doctors *incarnate* the destructive potential of science, leaving in its wake a world order in such disarray that it might make it impossible to reconstruct; in his biographical writings, engineers implicitly make use of science to rearrange the world, occasionally improving on the original "work," but more importantly and ostensibly, displaying a certain sensibility to the order of Creation. Inconsistent though it may seem at first sight, the figure of the scientist is throughout Stevenson's work a nexus where the individual is unmade and remade—and through him, society. Figures of science are also ways of figuring himself and his position in society, of course, but more importantly, though not unrelatedly, they embody forms of distribution of political power, denaturalizing the unceasing negotiation between society and the individual, the centre and the periphery, those with power and those who have none.

However, so often did Stevenson, an incurable patient and failed engineer, represent these types, and what they inherently meant, beyond their literal achievements or fiascos,[14] that he touched upon a figuration of science itself, not applied, as in the case of biology or industrial work, but more immaterial in nature, eluding simplistic personification. Energy and its circulation is a theme of paramount importance. The framework of thermodynamics is referred to as a way of metaphorically channeling "life energy," and in his anthropological vein,[15] a coupling of Darwinism and energy physics occurs: discourse on the past and future of the human civilization revolves around notions of conservation of energy or entropy, the "life-force" of mankind being either preserved or dissolved in the exercise of contemporary forms of imperial power.[16] Energy, its dissolution or circulation, become metaphors organically linked to the expression of politics in an abstract and universal sense, against the grain of Victorian elitism and racialism. This later desertion of role-models in Stevenson's use of science is striking in light of these times when he himself was facing a crisis concerning his own status as writer. While in Samoa, he tackled subjects not deemed particularly worthy of attention by the society he used to count in, and his social inscription not just in the British world, but also in the whole of the Western sphere of influence, his authority and authorship, came to be challenged, and he can be seen in various writings attempting to negotiate a new place for himself in an order he was made to feel he did not fit in anymore—a verdict he reluctantly agreed to. However, by examining modes of power attribution in the scientific and social fields, he does not just use and abuse science selfishly, as a field to borrow authority from, but as a way to restore this dignity that "Pulvis et Umbra" establishes should be found in all creatures and things, and perhaps even in inorganic matter.

Bibliography

Primary sources

Stevenson, Robert Louis. 1915. *Across The Plains With Other Memories And Essays*. London: Chatto &Windus.

—. 1925. *In the South Seas*. New York, Charles Scribner's Sons.

—. 1892. *A Footnote to History: Eight Years of Trouble in Samoa*. London: Cassell Company Limited.

—. 1901. "A plea for Gas Lamps" in *Virginibus Puerisque and Other Papers*. London: Chatto & Windus, pp. 271-8. [1878]

Secondary sources

Amossy, Ruth. 2006. *L'argumentation dans le discours*. Paris: Armand Colin.

Arata, Stephen. 1996. "The Sedulous Ape: Atavism, Professionalism, and Stevenson's *Jekyll and Hyde*" in S. Arata, *Fictions of Loss in the Victorian Fin de Siècle*. Cambridge: Cambridge University Press, pp. 33-53.

Besson, Cyril. 2014. "Le glaive contre le caducée: de l'indétermination de la figure du médecin chez Robert Louis Stevenson" in Lise Dumasy-Queffélec and Hélène Spengler (eds.), *Médecine, sciences de la vie et littérature en France et en Europe de la Révolution à nos jours*. Genève: Droz, pp. 259-60.

Cartwright, Nancy. 1983. *How the Laws of Physics Lie*. London: Oxford University Press.

Chassay, Jean-François. 2009. *Si la science m'était contée : des savants en littérature*. Paris: Editions du Seuil.

Cosslett, Tess. 1992. *The "Scientific Movement" and Victorian Literature*. Brighton: The Harvester's Press Limited.

Dryden, Linda, Stephen Arata and Eric Massie (eds.). *Stevenson and Conrad: Writers of Land and Sea*. To be consulted at: researchrepository.napier.ac.uk/.../Introduction_15_ (accessed December 6, 2015).

Gingras, Yves. *2010. Propos sur les sciences, entretiens avec Yanick Villedieu.* Paris: Raisons d'agir.

Harvie, Christopher. 1981. "The Politics of Stevenson," in Jenni Calder (ed.), *Stevenson and Victorian Scotland.* Edinburgh: Edinburgh University Press, pp. 107-125.

Jolly, Roslyn. 2009. *Robert Louis Stevenson in the Pacific: Travel, Empire and the Author's Profession*. Farnham, England: Ashgate Publishing Ltd.

Jordanova, Ludmilla. 1994. "Melancholy Reflection: Constructing an Identity for Unveilers of Nature", in Stephen Bann (ed.), *Frankenstein, Creation and Monstrosity*. London, Reaktion Books.

Kuhn, Thomas S. 1957. *The Copernican Revolution: Planetary Astronomy in the Development of Western Thought*. Harvard: Harvard University Press.

—. 1962. *The Structure of scientific Revolutions*. Chicago: The University of Chicago Press.

Lepeltier, T. (ed.). 2013. *Histoire et philosophie des sciences*. Auxerre: Sciences Humaines Editions.

Lucas, John. 1971. "Conservatism and Revolution in the 1880's" in John

Lucas (ed.), *Literature and Politics in the Nineteenth Century*. London: Methuen and Co. LTD, pp. 173-219.

MacDuffie, Allen. 2006. "Irreversible Transformations: Robert Louis Stevenson's *Dr. Jekyll and Mr. Hyde* and Scottish Energy Science", in *Representations* 96, Autumn, pp. 1-20.

Reid, Julia. 2006. *Robert Louis Stevenson, Science, and the Fin de Siècle*. Basingstokes: Palgrave Macmillan.

Snow, Charles. 1964. *The Two Cultures and A Second Look*. Cambridge: Cambridge University Press.

Stiles Anne. 2006. "Robert Louis Stevenson's *Jekyll and Hyde* and the Double Brain" in *SEL* 46, 4, Autumn, pp. 879-900.

Sussman, Herbert. 1968. *Victorians and the Machine, The Literary Response to Technology*. Cambridge, Mass.: Harvard University Press.

Notes

[1] A good example, although far more radical than anything Stevenson experienced, is the case of Arthur Conan Doyle and Sherlock Holmes.

[2] For this quotation and all the others from that book, my translation.

[3] See R. K. Merton's now classic presentation of the four norms of science, aptly summarized in T. Lepeltier (ed.), *Histoire et philosophie des sciences*; Auxerre, Sciences Humaines Editions 2013, p. 182. This vision leads not just to a polity, but a politics of sorts (in the framework of a republic of the sciences), governed by an ethos (code of ethics) that leads to a "liberal, egalitarian and democratic" perception of the scientific field (*ibid*), the guarantee that it will not be infringed upon by the political or economic fields. This of course is not about the contents of the sciences themselves, but about their *production*. This somewhat angelic view was later contested by the notion that science is a market like any other, where only the nature of the "goods" exchanged differ: in the scientific community, knowledge and peer recognition are substituted to goods and services (Lepeltier 183).

[4] "Normal science" is the way in which Kuhn calls the exercise of science during periods when there are no great controversies in the field.

[5] *Logos* (discourse as demonstration), *ethos* (the moral character of the speaker as "projected" by him into his performance), and *pathos* (the strategies through which emotion can be brought about in the receiver).

[6] In *Across the Plains* (1915), pages 112-125. To be consulted at http://www.gutenberg.org/files/614/614-h/614-h.htm. Accessed on January 13, 2016.

[7] It has not been possible at the time of writing to lay hands on a copy of the book. All quotations are taken from Gutenberg.org: http://www.gutenberg. org/files/ 280/280-h/280-h.htm (accessed Dec. 14, 2015).

[8] In the sense meant in Roman antiquity, of "the forms of the family's, or clan's, power of continuing itself by reproduction, which were temporarily in keeping of

the heads of the family or clan and passed at death to their successors." *Encyclopedia Britannica*, 1973 edition, Micropedia, Vol VI p. 467.
[9] At least, this "romance of engineering" is what Stevenson chooses to focus on in the bulk of his evocation of the profession. The reality of the drudgery of the other part the job, sitting at the drawing board, and its deleterious consequences on the engineer, is only mentioned in passing, but casts far less enticing a light on the vocation: as he probably imagined himself doing, one may go mad, or at least fall prey to schizophrenia if one's character is not strong enough. See http://www.gutenberg.org/files/30990/30990-h/30990-h.htm#page62, page 174.
[10] On patents, see Yves Gingras, *op. cit.,* p. 64. The institution of patenting is a also form of peer-recognition for not just one's skill as inventor, but in terms of one's social responsibility. By renouncing this respect from their peers, the Stevensons are shown to take the greatest possible sacrifice for engineers. It should be noted that, whether "innovative" or not, all of Robert Louis's books bear the name of their author, a betrayal of the family inheritance, surely? To quote from chapter II of *Records*..., "engineering looks one way, and literature another."
[11] Stevenson treats the "literary" claims of his grandfather differently, rebuking him for his "monuments of misdirected literary energy, which leave upon the mind of the reader no effect but that of a multiplicity of words and the suggested vignette of a lusty old gentleman scrambling among tangle."
[12] To quote Sussman on this point: "the very concept of a coherent natural order is a construct of the scientific Imagination. The scientist sees what he has chosen to see: his choice to see 'objective' facts and rational law is based on his belief in their value. The scientific world view is both solidly based on Nature's truth and law; and precariously based on human choice, a conscious limitation to certain ways of seeing Nature. This implies that other ways of seeing, other choices of symbols, might be equally valid. Here the scientists would reply that their way of seeing is preferable because of its results [...]". Herbert L. Sussman, *Victorians and the Machine, The Literary Response to Technology*, 35.
[13] This tends to mitigate the optimism with which someone like Charles P. Snow sees the scientist's mindset, characteristic of the vocation as social force: "Most of the scientists I have known well have felt [...] that the individual condition of each of us is tragic. [...] Each of us is solitary, each of us dies alone: all right, that's a fate against which we can't struggle-but there is plenty in our condition which is not fate, and against which we are less than human unless we do struggle. [...] They [scientists as a group] are inclined to be impatient to see if something can be done: and inclined to think that it can be done, until it's proved otherwise. That is their real optimism, and it's an optimism that the rest of us badly need". *The Two Cultures and A Second Look*, pp. 6-7.
[14] See when he ever so fleetingly evokes his father's failure with the harbour of Wick, in a passage so apocalyptic and overstated that it becomes bombastic: "the harbour of Wick, the chief disaster of my father's life, was a failure; the sea proved too strong for man's arts; and after expedients hitherto unthought of, and on a scale hyper-cyclopean, the work must be deserted, and now stands a ruin in that bleak, God-forsaken bay", p. 132.

[15] Especially in his Samoan writings. *In the South Seas* and *A Footnote to History: Eight Years of Trouble in Samoa*.

[16] See Roslyn Jolly, pp. 29-97 especially, and Julia Reid, pp. 138-173.

PART II:

THE MIGRATION OF SCIENTIFIC CONCEPTS

ANDREW CARNEGIE AND MANAGERIAL DARWINISM: FROM SCIENTIFIC DISCOURSE TO MANAGEMENT CONCEPTS AT THE TURN OF THE 20TH CENTURY

CHRISTIAN LEBLOND

> "The millionaire class needs no defense",
> —Andrew Carnegie, *The Gospel of Wealth*, 77.

Andrew Carnegie (1835-1919) is a well-known figure, an industrialist turned philanthropist in later life. He is not always perceived in a positive light and many regard him as the epitome—almost the caricature—of the Robber Baron of the Gilded Age. The ruthless builder of an industrial empire in an era of non-regulated capitalistic development, Carnegie could easily be dismissed as a purely opportunistic predator whose sense of ethics was primitive. This chapter, however, will try to nuance this simplistic approach by assessing contextual elements showing the complexity of this figure, the influences that shaped his vision, and the relevance of his writings on management. Andrew Carnegie was a successful manager but, in his effort to reconcile his vision of management with ethics, he reveals his most intriguing and perhaps most interesting facet as a long-forgotten post-political Darwinian.

Andrew Carnegie was mainly a man of action, not a prolific author or an academic. Yet he probably deserves a chapter in the history of economic thought. Building on the scientific discourse of Darwin and its subsequent transposition to human societies by Herbert Spencer, Andrew Carnegie turned his own experience at the helm of his industrial empire into raw materials for a reflection on the new form of human organization he knew best: the corporate world. Carnegie therefore provides the missing link in the history of ideas between social Darwinism and the intellectual apparatus that proved necessary to manage giant corporations that American industrialists constituted at end of the 19th century. Through his explicitly Darwinian vision of management, he thus contributed to the emergence of management concepts in the United States at the turn of the 20th century, and paved the way for subsequent political models directly influenced by new forms of corporate governance. Andrew Carnegie therefore contributed to the gradual migration of scientific concepts based on the observation of species to the field of human governance and political discourse.

His main work in this field is a famous collection of texts, initially newspaper columns,[1] published in 1889 in the USA under the collective title of *Wealth* and later in the USA and Great-Britain under the title *The Gospel of Wealth* (1901).[2] He had no difficulty writing on the matter with authority since he was then considered the wealthiest man in the United States as the founder and manager of an industrial empire based on steel and railroads. In 1901, the Carnegie Steel Corporation, which was a large vertically-integrated company,[3] was taken over by a financial consortium led by New York banker J. P. Morgan.[4] Under its new name and new management, U.S. Steel moved on and became the largest steel corporation in the world and remained on top for several decades.

Political Positioning

Andrew Carnegie admits he is intellectually indebted to previous thinkers and theorists, that he himself identifies as Darwin and Spencer. Nevertheless his family background, the episodes of his professional ascent and, above all, his personal experience as an investor and empire builder may have played an even more defining role. Perhaps the main interest of Carnegie's writings is less the originality of his views than the dual source of his vision, as an avid reader and ideologue as well as a man of action characterized by the proverbial can-do attitude of American entrepreneurs.

Andrew Carnegie went down in history as the triumphant self-made man who rose from the proverbial "rags to riches", who achieved a very high social status from humble beginnings, thus illustrating the theme of America as the land of opportunity. Yet, young Carnegie, poor but educated[5], did not qualify as a working class hero. The Carnegie family was not affluent, but its members were weavers by trade, which meant that, in the "cottage" system that was still widespread in Scotland, they were actually in an intermediate position—the equivalent of present-day subcontractors. The Carnegie family was therefore financially on a par with the working class, but ideologically, the family narrative may have highlighted enterprise rather than exploitation.

Carnegie also embodied the old school of entrepreneurs, who finally had to concede defeat under the pressure of a new breed of financiers. The take-over of Pittsburgh smokestacks by New York banks, the victory of patrician banker J. P. Morgan over Carnegie, was a crucial episode in the first financialization of the US economy. Strikingly, Carnegie embraced the part of apologist of a system that had just made him extremely wealthy as an individual, but redundant as a manager.

Initially, Carnegie was perceived as a progressive for his action in favor of Filipino independence. He famously offered a huge portion of his fortune to the people of the Philippines so it could buy its independence and not become an American possession, which would sound like a generous anti-imperialist gesture if we overlooked comments like:

> What we have to face is the question whether we should embark upon the difficult and dangerous policy of undertaking the government of alien races in lands where it is impossible for our own race to be produced. (GW, 155)

His reticence was apparently motivated by biology, the Philippines being a region of the globe that had "produced" an ethnic group that was "alien" and therefore impossible to assimilate.[6] A man of his time, Carnegie seemed to have been instinctively Darwinian in his approach to what constituted the appropriate space for what he considered his own ethnic group to expand. At the end of a long process of natural selection and adaptation, the environment of the Philippines had shaped the local ethnic identity, which was very different from the Americans, so trying to settle in that region of the globe was not condemned by Carnegie as an act of imperialist aggression but as biological nonsense. The ambiguity about Andrew Carnegie's acceptance of progressive ideas also derives from a number of statements that may sound generous, for instance concerning wages:

> Millionaires make no money when compelled to pay low wages. Their profits accrue in periods when wages are high, and the higher the wages that have to be paid, the higher the revenues of the employer. It is true, and not false, therefore, that capital and labor are allies and not antagonistic forces, and that one cannot prosper when the other does not. (GW, 53)

This very statement, as well as Carnegie's defense of labor unions in *Forum Magazine* (1886), was contradicted by Carnegie's subsistence wage policy and brutal methods, especially the shootout between strikers and Pinkerton detectives hired by his company at the Homestead plant in 1892. So the only correct interpretation of these lines is that productivity lifts both wages and profits, not that profits should be reduced to increase the purchasing power of workers who may not subscribe to their employer's vibrant defense of capitalism:

> Consider the change, nay, the revolution! Now the poorest laborer in America or in England, or indeed throughout the civilized world, who can handle a pick or a shovel, stands upon equal terms with the purchaser of his labor. (GW, 108)

For the author of these lines, a free and unregulated labor market is where supply harmoniously meets demand, where employers and laborers make mutually beneficial transactions on an equal footing—a view that is not unanimously shared. Institutional and Marxist economists consider that selling one's labor on the market to make a living automatically places the laborer in a situation of dependency and exploitation. The interest of these statements is not really in their content—a form of market optimism which is easily decoded ideologically—but the total absence of provisions for any regulating body, a role traditionally assigned to the State and its extensions. To a certain extent we could say that Carnegie's discourse is entirely devoted to economic mechanisms precisely because, as a follower of Herbert Spencer, he considers that the irrelevance of the State has already been amply demonstrated.

To begin with, H. Spencer's definition of the ideal state is both minimal and optional:

> What is it that we call the state? Men politically associated. How associated? Voluntarily. For what purpose? For mutual protection. Men voluntarily associated for mutual protection: this then is our definition. (*Social Statics*, 172)

Spencer's demonstration is impeccably logical: if the state is a reunion of individuals who opt for a sort of collective mutual protection scheme, opting out should be a possibility:

> If every man has freedom to do all that he wills, provided he infringes not the equal freedom of any other man, then he is free to drop connection with the state—to relinquish its protection, and to refuse paying towards its support […] He cannot be coerced into political combination without a breach of the law of equal freedom; he can withdraw from it without committing any such breach; and he has therefore a right so to withdraw. (*Social Statics*, 131)

Trying to extract some sort of political doctrine from Carnegie's writings, therefore, is bound to prove extremely frustrating, because his reflection starts from where Herbert Spencer had left the issue, asserting the right to political separatism, in an extreme form of individualism heralding the Libertarians of the 1970s. This does not mean that Carnegie did not have an opinion on human governance but that we can only get a passing glimpse of it especially when he refers to his mentor:

> Herbert Spencer's great law will be further vindicated: as power is held arbitrarily by king or chief, the military type is developed, and wars of dynasties and aggression ensue. As power passes to the people the industrial type is developed, and peace ensues. (GW, 217)

In a few words that we can only decode in relation with Spencer's ideology, Carnegie sketches a sort of ideal society in which "the industrial type" embodies the most democratic and peaceful form of citizen as well as economic agent–the two being virtually impossible to distinguish. In other words, the debate on the best form of government having been shown to be irrelevant, Carnegie will focus on the crucial issue of industrial governance and on his capacity to influence his fellow-Americans by spending his vast fortune for the promotion of his ideas, contributing to what he sees as a collective progress. In that very particular sense, therefore, Carnegie was and remained a committed progressive.

In the second phase of his life, Carnegie devoted himself entirely to philanthropic pursuits, and went on distributing his wealth to countless public libraries and concert halls–as he had done for the famous Carnegie Hall inaugurated in 1891. Here, the influence of his early years was obvious, since he had used the resources of the public libraries open to the general public free of charge to gain access to literary and scientific culture.

So, Carnegie was intellectually a disciple of Charles Darwin and Herbert Spencer, and his main merit may be his methodical transposition of evolutionary and Spencerian concepts to corporate management. Carnegie saw the potential of this new and controversial scientific discourse, and used it in the description and theoretical justification[7] of his business practices. Carnegie was instrumental in the emergence of a whole new branch of economics centered on the firm rather than the economic system as a whole.[8] Yet, Carnegie's complex personality is not entirely accounted for by this type of analysis and this icon of the Gilded Age turned ideologue and philanthropist remains a paradoxical figure on other accounts as well.

Groundwork for Managerial Theory

An influential columnist, Carnegie publicized his views on the virtues of an unregulated market economy. At a time when the US was undergoing what W. W. Rostow called an economic take-off,[9] Carnegie's testimony is interesting because he was both a player and a commentator in the momentous transformations that allowed his country to turn into a major economic power. Since Carnegie Steel was expanding in this very particular context by conquering new markets and by taking over competitors, Carnegie was probably convinced that the dynamics of his ascent unequivocally demonstrated his expertise and capacity for leadership.

His main field of interest obviously lay in corporate economics and management, which means that he was not—technically speaking—a political scientist and barely an economist.[10] Yet, he spent most of his early life making investment decisions and managing increasingly complex companies. In the wake of Herbert Spencer, who denounced the British Parliament and "the immense positive evils entailed by over-legislation" in 1853 (*Over-Legislation*, 7.42), Carnegie saw in private companies an emerging type of organization gradually pushing politics and policy into irrelevance because of its unprecedented aptitude to generate affluence and stability. By allowing the search for ever-increasing efficiency to be carried out unrestricted, Carnegie ultimately aims at reaching a stage of history where the new class of rational managers like himself would take over from well-meaning but ineffective politicians.

Carnegie had to lift all mental restrictions to pursue this goal methodically by demonstrating that the social utility and responsibility of business enterprises was strictly equal to their economic efficiency. By insisting that efficiency was in fact ethical, he was preparing both public and academic opinion for the notion that the quest for efficient management techniques should not be restricted by traditional social conventions, and that research in that field should explore all the possible perspectives because the end result was bound to be positive for society as a whole. In that sense, by stressing the ethical—albeit paradoxical—dimension of the entrepreneur's actions, Carnegie cleared the way for the invention of scientific management. Furthermore, as increasingly well-managed firms are bound to take precedence over the State as a form of human organization, so political thought, long the realm of purely speculative philosophers, must give way to other kinds of intellectual speculation. The new breed of thinkers, in Carnegie's wake, must focus on fitting in with the mechanisms of self-regulating markets, considered as a spontaneous and irrevocable order of things and not as a man-made system. Carnegie's economic discourse can be termed anti-political or post-political, insofar as it envisions the withering away of the state, a characteristic he shares with many die-hard free-marketeers up to present-day libertarians.[11] He actually goes further than Herbert Spencer's opting-out provision on the capital point of collective security:

> Rioters assembling in numbers and marching to pillage will be remorselessly shot down; not by the order of a government above the people, not by overwhelming standing armies, not by troops brought from a distance, but by the masses of peaceable and orderly citizens of all classes in their own community. (GW, 143)

One is impressed by the almost Proudhonian nature of this vision of an orderly society where crime is suppressed by a free and spontaneous surge of law-abiding citizens without the need for riot police or National Guardsmen to be sent in.[12]

So what definitely connects Carnegie's writings with the scientific discourse of his time is the constant reference to Darwin and Spencer's social darwinism. This in itself incites us to review the standing of Carnegie as a long-forgotten post-political Darwinian, whose talent was to transpose Darwin's concepts to explain and even to justify the necessary actions of the rational manager, who may resort to very rough tactics against his competitors as well as his workforce without feeling ethically destabilized.

From Darwinism to Corporate Structures

In the opening pages of *The Origin of Species*, Malthus is hailed by Darwin as one of his major sources of inspiration[13] and the influence is obvious when, starting from the fact that living creatures are endowed with "geometrical powers of increase", he reasons that *"many more individuals of each species are born than can possibly survive" so the survivors are sorted out by a process of natural selection which "almost inevitably causes much Extinction of the less improved forms of life."*[14] If the reader replaces "individual" by "firms" the basic element of Carnegie's justification of strict management in a competitive business environment is pretty clear. The sorting out process needed to separate the viable forms of business venture from the others is a transposition of Darwin's original formulation: "Natural selection almost inevitably causes much natural extinction of the less improved forms of life" may be applied to the competing firms' "struggle for existence." Competition plays the same role as natural selection by eliminating the less improved forms of management–to paraphrase Darwin, which means that sooner or later, less talented managers run their firms into bankruptcy.

This transposition is relatively easy to figure out, but some implications are less immediately perceptible. For instance, according to the theory of evolution, species were not created independently, but current species are in fact “lineal descendants of some other and generally extinct species”–which means, when the rule is applied to corporate management, that the law of competition is generalized, and that two companies even with very different core activities are competitors rather than partners.

On the whole, the influence of Darwin seems to have gone along a rather circuitous route before reaching Andrew Carnegie. Herbert Spencer carried out a first transfer of evolutionary concepts from biology to the field of human societies–the result being Social Darwinism which, later on, was to provide Carnegie with the conceptual materials for his reflection leading eventually to the emergence of scientific management. Very often, both influences are so closely intertwined as to become impossible to trace back to their original author, for instance when *Carnegie comments on the law of competition:* "while the law may be sometimes hard for the individual, it is best for the race, because it insures the survival of the fittest in every department (GW, 3). In that particular case, biological evolutionary discourse seems to function metaphorically to convey notions actually connected to the realities of competing businesses. On the whole, though, the influence exerted by Spencer seems to prevail over all others. This observation, derived from carefully reading Carnegie's texts, is confirmed by compelling evidence from biographical details. Carnegie's *autobiography was published posthumously in 1920* under the supervision of his wife. Among the collected papers, the reader finds an eight page chapter entitled “Herbert Spencer and His Disciple”–leaving no possible doubt about the late Carnegie's intellectual debt to the British philosopher.

Scientific Management and Ethics

It would be slightly excessive to proclaim that Andrew Carnegie is the missing link between Darwin and the so-called evolutionary economists.[15] However, it seems plausible to place him somewhere between the Spencerian school of social sciences and subsequent theoretical work on the firm and the market. By this we mean economists like Thorstein Veblen, an institutionalist, who was actually Carnegie's contemporary.[16] At the other end of the political spectrum we must quote the Austrian school, led by Joseph Schumpeter whose *Theory of Economic Development* was published as early as 1911–but translated into English much later in 1934.

Here, Carnegie is indebted to Darwinian biology through Herbert Spencer, whose economic vision equates the inner mechanisms of the firm to those of a living organism, not a man-made machine:

> Between creatures of the lowest type, and creatures of the highest, we similarly find the essential difference to be, that in the one the vital actions are carried on by a few simple agents, whilst in the other the vital actions are severally decomposed into their component parts, and each of these parts has an agent to itself. In organizations of another order the same principle is apparent. When the manufacturer discovered that by confining each of his employés [sic] wholly to one process, he could immensely increase the productive powers of his establishment, he did but act upon this same rule, of one function to one organ. (*Social Statics*, "The limits of State duty", 171)

This biological metaphor reaches far beyond rhetorical inventiveness, because this type of organization has to be *discovered* experimentally, but not empirically because the biological model "one function to one organ" actually provides a blueprint for the discovery of the optimum outlay of workers and resources. Applying these management principles–allegedly inspired by functional biology–cannot be regarded as morally debatable, and guilt is totally evacuated from the decision-making process.

As a good economic liberal, Carnegie's intention was to reconcile Ethics, optimism and progress with the manifest ruthlessness of the law of competition and exploitation of the workforce:

> The laws of accumulation will be left free, the laws of distribution free. Individualism will continue, but the millionaire will be but a trustee for the poor, entrusted for a season with a great part of the increased wealth of the community, but administering it for the community far better than it could or would have done for itself. (GW, 18)

In Marxist terms, what is implied here is that the confiscation of the added value by the entrepreneur is legitimate as long as he acts as "a trustee for the poor", a sort of fund manager who is temporarily in charge of poor people's money because he is more competent and is likely to make better use of it. On the other hand, a reformulation in Weberian terms would probably lead the reader to the puritan notion of the Stewardship of riches,[17] which rests on the assumption that wealth is totally devoid of intrinsic value and serves only a sign of election.

In the same line, Carnegie defends the idea that "lives of poverty and struggle are advantageous" (*GW*, 35) because individuals submitted to the pressure of hardship will find the resources to perform better and contribute to progress. On the whole, the kind of evolutionary optimism professed by Carnegie is rooted in the long term certainty that "the race toils slowly upward step by step". His evolutionary conception of the firm is indeed a watered-down version of the struggle for life because the losers are not put to death. On the contrary, the production units and the workers are taken over by the new management, and the best practices end up as the universal rule of management, contributing to an optimization of resources and eventually to the general welfare of the economy. Consequently, nobody will shed a tear for the "small, petty master in his little store" who is supplanted by the "much more important manager of a department" (*GW*, 94), even though the latter is dependent on the senior managers of the firm he has joined. On the whole, the destruction of small businesses is presented in a very positive light, which leads us to the conclusion that Carnegie may have opened the way for the approach of the Austrian school. Four decades later, Schumpeter published *Capitalism, Socialism and Democracy* in which the whole of chapter 6, famously titled "The process of creative destruction", seems to give a final formulation to Carnegie's intuitions on the virtues of economic competition formulated in Darwinian terms like "natural selection" or "survival of the fittest". But the convergence between Carnegie's approach and the Austrian school may go much further, if we keep in mind that among the three factors for economic development identified by Schumpeter in 1911[18] the entrepreneur figures prominently as the key person who perceives the business opportunities of recent disruptive innovations and reorganizes his operations for the production of new goods or services. Andrew Carnegie was not the only inspiring figure of the American industrial revolution, but the classic Schumpeterian definition of the entrepreneur almost seems to be modeled on him or his likes. Perhaps even more deeply, the methodology of Austrian economists—Praxeology—may owe some of its inductive approach to the way men of action like Carnegie tried to give a rational account of their managing practices:

> Praxeology rests on the fundamental axiom that individual human beings act, that is, on the primordial fact that individuals engage in conscious actions toward chosen goals. This concept of action contrasts to purely reflexive, or knee-jerk, behavior, which is not directed toward goals. The praxeological method spins out by verbal deduction the logical implications of that primordial fact. In short, praxeological economics is the structure of logical implications of the fact that individuals act. (Rothbard, "Praxeology", 75)[19]

Needless to say, Marxists like Werner Sombart very quickly explored another line of reasoning, and concluded that the destruction of competitors actually resulted in the concentration of wealth in the hands of an ever-shrinking elite, a situation known as *monopolistic capitalism*.

This presentation of Carnegie's intellectual influence would not be complete without mentioning the fact that he paved the way for the emergence of management as an autonomous branch of economics around the time *The Gospel of Wealth* was published. Students of the history of economic ideas have to admit that the spirit of the time, the *zeitgeist*, was obviously very favorable to a new breed of hands-on micro-economists, as opposed to ideologically motivated macroeconomists. Carnegie, both an observer and a player[20] is in the same league as Frederick Winslow Taylor, who gave his name to Taylorism, and Henri Fayol, generally regarded as the other founder of scientific management. Taylor was working as an engineer for Bethleem Steel in 1901 and published *The Principles of Scientific Management* in 1911, whereas Fayol, a graduate of the *Ecoles des Mines de Saint Etienne*, published *Administration industrielle et générale* five years later (1916). What they have in common with Carnegie is a hitherto uncommon interest in the role of the managers, whose function is to concentrate enough power[21] to enforce standardization and procedures, turning the factory into a technocracy for efficiency's sake.

Carnegie was instrumental in translating Spencer into the corporate logic, but ironically, we may wonder whether this transposition backfired. Indeed, the technocratic model of an efficient and ever expanding company under the rational leadership of its managers may have become the vehicle of choice for human societies as a whole and not only business enterprises. Seen from this particular perspective, corporate governance and the art of governing human societies may have interacted in more than one direction in the first part of the 20th century. The imperfections of political organizations denounced by Spencer may have inspired the quest for an efficient type of organization to turn to the field of industry. Yet subsequent forms of civil governments, especially authoritarian ones, may be indebted to the forms of centralized management initially developed for industrial organizations.

Conclusion: from Rags to Riches … to Leisure Class?

Throughout this chapter, we have insisted that Carnegie was in a category of his own, and that he stood out in the history of American economic thought as an entrepreneur and a thinker. We have already underlined that he was a near perfect illustration of the American Dream, who succeeded through sobriety, hard work, shrewd investments and risk-taking–in a nutshell the cardinal virtues of the Calvinist entrepreneur as depicted by Max Weber even though this scion of a Calvinist dynasty was in fact an atheist. Actually, *The Gospel of Wealth* is far from being a pure pragmatic analysis. This collection of texts is many other things, including possibly the first American book on self-improvement, exhorting his fellow Americans to follow his inspirational teaching. Carnegie's writings have an obvious dimension of self-exploration, self-vindication or even self-celebration, to such an extent that *Wealth* can hardly be read independently from his *Autobiography*. Obviously, the motives behind this *pro domo* discourse are not purely speculative and can be construed as an attempt at rationalization in the psychoanalytical sense of the term:

> A process by which a subject aims at giving a coherent explanation from the logical point of view, or acceptable from the moral viewpoint, to an attitude, an action, an idea, a feeling etc..., the genuine motives of which are not perceived. (Laplanche and Pontalis, 387)[22]

It is clear from the start that Carnegie is not a neutral outside observer since he looks back at his own career as the CEO of Carnegie Steel. If we subscribe to this interpretation, Carnegie may be deluding himself, and his readers, by producing a discourse whose primary function is not to lift the veil on a ground-breaking managerial experiment but to put an acceptable front on subliminal motivations and urges. For US economist Thorstein Veblen the extreme diversity of norms and values can only be accounted for by cultural factors, some of them very archaic, which are deeply rooted in the past and may become unconscious but cannot be erased. Veblen considers that economics is not an evolutionary science, since archaic patterns will almost inevitably surface sooner or later. When a Robber Baron turns into a member of the leisure class, and adopts the lifestyle of the aristocrats, he certainly conforms to Veblen's model of the gentrified predator who gains social prestige after reinventing himself.[23] So Carnegie's writings and ideology have a manifest Darwinian content, but the key to Carnegie's psyche may be Veblen's analysis. At the end of the day, we have good reasons to wonder whether this impeccably Darwinian manager outlived himself after being bought out by JP Morgan. Ironically, Andrew Carnegie, especially in old age, had apparently become gradually more veblenian and far less Darwinian.

Bibliography

Carnegie, Andrew. 1963. *Autobiography of Andrew Carnegie*. NY: Pyramid. [1920]

—. 1901. *The Gospel of Wealth and Other Timely Essays*. New York: Century.

Fayol, Henri. 1918. “Administration industrielle et générale”, *Bulletin de*

la Société de l'Industrie Minérale, N° 10, 5-164, 1916. [reprints by Dunod after 1918]
Laplanche, Jean et Jean-Bertrand Pontalis. 1968. *Vocabulaire de la psychanalyse*. Presses univ. de France. [Translated into English as *The language of psychoanalysis*. Karnac Books, 1988]
Rostow, Walt Whitman. 1960. *The Stages of Economic Growth: A non-Communist Manifesto*. Cambridge: CUP.
Rothbard, Murray N. 1976. "Praxeology: the Methodology of Austrian Economics" in Edwin G. Dolan, ed. *The Foundations of Modern Austrian Economics.* Kansas City, Sheed and Ward, 1976, b75-88. To be consulted at: http://oll.libertyfund.org/titles/104 (accessed 22/05/2015).
Schumpeter, Joseph, Alois. 2011. *The Theory of Economic Development: An Inquiry into Profits, Capital, Credit Interst and the Business Cycle*, Transaction (translated from the German in 1934). [German edition 1911]
—. 2012. *Capitalism, Socialism, and Democracy.* Start. [1942]
Sombart, Werner. 2013. *Krieg und Kapitalismus*. BoD. [1913]
Spencer, Herbert. 1851. *Social Statics or The Conditions essential to Happiness specified, and the First of them developped*. London, John Chapman. Available on the Online Library of Liberty website < http://oll.libertyfund.org>
—. "Over-Legislation". 1992. *The Westminster Review for July, 1853.* Reprinted in *The Man Versus The State, with Six Essays on Government, Society, and Freedom*. Econlib site, p. 7.42 To be consulted at: http://www.econlib.org/library/LFBooks/Spencer/spnMvS5.html (accessed 22/05/2015).
Taylor, Frederik Winslow. 1911. *The Principles of Scientific Management,* New York, Harpers and Brothers.
Veblen, Thorstein. 1978. *The Theory of Business Enterprise*. Transaction Publishers. [1899]
—. *The theory of the leisure class; an economic study of institutions*. 2005. Aakar Books.
—. 1898. "Why is economics not an evolutionary science?" *The Quarterly Journal of Economics*, 1898, vol. 12, no 4, pp. 373-397.

Notes

[1] The articles in *Wealth* range from 1886 to 1900 and have copyrights showing as many as six different publishers.
[2] All the references are to this New York Century edition of 1901. Subsequently *GW.*
[3] Technically a holding company, even though the traditional and inaccurate term *trust* seems generally accepted.
[4] In fact, Carnegie sold his operations to J. P. Morgan for $480m, which made him—ironically—the richest man in the US precisely when he left his corporate empire.
[5] Carnegie had little formal education but was a voracious reader, and was mainly self-taught.
[6] Similar arguments had been used about Mexicans to limit the annexation of territories to unpopulated or scarcely populated areas in the West as a result of the treaty of Guadalupe Hidalgo.
[7] By describing them as *inevitable.* This is not an endorsement of this extremely ideological justification—simply a statement of Carnegie's main objective.
[8] That is. microeconomics as opposed to macroeconomics.
[9] In his 1960 classic study titled *The Stages of Economic Growth,* Rostow distinguished five different stages.
[10] The term *economist* remained rather imprecise as economics gradually emerged as an autonomous branch of learning, different from philosophy and law in the early 20th century.
[11] The best example is probably US economist Murray Rothbard.
[12] In the infamous shootout at the Homestead plant in 1892, Pinkerton detectives were not "*masses of peaceable and orderly citizens of all classes" but worked for a* private security company who had more employees on its payroll than the US army. Driven out by armed strikers, they eventually retreated and the US military was finally sent in.
[13] A very long list of previous thinkers that Darwin himself finds influential can be found in a footnote of the opening chapter of the second edition of *The Origin of Species* (January 1860). It includes Malthus.
[14] Introduction to *The Origins of Species*, first edition, 1859. Chapter 4 is entirely devoted to the analysis of this natural selection process.
[15] The term of evolutionary economics was coined by Veblen in an 1898 article (in which he refuted the notion) but numerous economists have described themselves as "evolutionary", among them Armen Alchian and Kenneth Boulding, a list to which some specialists also add Kenneth Arrow, Ronale Coase and Douglas North, three famous Nobel prize winners.
[16] In the conclusion, the reader shall see how Veblen's concept of the Leisure Class can shed light on Carnegie's final years–but in this paragraph the allusion is to Veblen's subsequent theory of business enterprise.
[17] The term *stewardship* used by Paul (1:25)–also in Ephesians for example 1:10—interests economists as well as students of Puritan theology because it is

used to translate the Greek word *oikonomia*—i.e. *dispensation of wealth* (God dispenses riches to the Elect).

[18] Three characteristics are generally listed: the endogenous nature of the process, the key role of the entrepreneur and the availability of venture capital.

[19] The paragraph quoted is signed by Murray Rothbard, who specifies that the idea comes from Ludwig Von Mises.

[20] Admittedly, these roles were largely consecutive.

[21] Among Fayol's famous Fourteen Principles of Management, at least 8 focus on power: authority, discipline, unity of command, unity of direction, subordination, centralization, hierarchy, order.

[22] Translation is mine. Rationalization is not considered by psychoanalysts a defense mechanism but only a secondary elaboration.

[23] Carnegie bought a dilapidated estate in Scotland around Skibo Castle. His intention was to end his life and be buried on the grounds, a plan that the First World War made impossible to execute. Carnegie's tombstone, though, was brought to the USA from Skibo.

'EQUIVOCATING UPON LAW AND SCIENCE': SCIENCE, POLITICS, AND THE MIGRATION OF HUMANITARIANISM IN NINETEENTH-CENTURY AMERICA

JEFFREY MULLINS

During the summer of 1846, two of America's leading politicians met in a remote county courthouse in central New York to battle over a seemingly open-and-shut case. When William Freeman killed four members of the Van Nest family in March of that year, and soon admitted doing so to police officials, there was little indication that a months-long legal battle would ensue, or that it would bring politics and medical science into the courtroom or splashed across newspapers nationwide. When William Seward (the ex-Governor of New York, and soon to be Abraham Lincoln's Secretary of State) took on Freeman's cause in the name of "science" and "humanitarianism," however, political adversaries of this prominent Whig Party leader felt the need to counter this move. Thus Democratic Party leaders sent John Van Buren (New York's Attorney General, and the son of the former U.S. President) to take up the other side of the case. The immediate issue was the possibility that Freeman might not be mentally competent, but much larger issues concerning morality and scientific knowledge—with all of their political and social implications—were at stake.

At the very core of the Freeman case was a question that many Americans were increasingly asking themselves: could science pose a threat to democracy? For nearly three generations after the founding of the United States, to the extent that they thought about it at all, Americans had generally seen science as the savior of the republic. From agriculture to engineering, medicine to morality, a scientific approach to the problems facing the new nation promised to help Americans prosper, and to do so based on an extension of the very Enlightenment principles upon which the country was originally founded. Conventional narratives of science in nineteenth-century America generally depict science as being on an ascending (if bumpy) arc to legitimacy and eventual dominance. A closer look at the evidence, however, reveals that science—especially as applied to human beings—suffered a severe setback beginning in the 1830s. The causes of this setback lay less in the dynamics of science itself and more in the political tensions of the era. The development that triggered the reversal in attitudes was a shift in perceptions of humanitarianism and social reform, leading to a reappraisal of the scientific model of human nature that provided the conceptual foundation for such endeavors.

The approach of this essay stands in contrast to the majority of existing scholarship on both politics and the sciences (medical and otherwise) in antebellum America. Political historians have mainly concerned themselves with either the process of party formation and the contests between them, or with the development of the modern nation state. Historians of medicine and science in this era have largely focused on the technical developments of a particular field, the process of professionalization, or the culture internal to the practitioners of a given field. As will quickly become clear, this essay takes all of this important scholarship as its immediate context. Building on this prior work, the goal here is to look at how both politics and medical science operated in contexts beyond those that we normally associate with them (e.g. elections, legislative debates, the physician's office, medical dialogues). By focusing on how these areas of endeavor played out in actual (and broader) practice, we can gain some insights not usually possible with more conventional approaches. Scholars such as policy historian William Novak, "emphasiz[ing] the actual day-to-day *conduct* of governance," instead of the typical focus on policy statements and party platforms, have managed to fundamentally re-orient our understanding of the nineteenth-century American polity (Novak, 8). In the realm of science, Katherine Pandora has recently called for a move away "from studying scientists *in situ* to the pursuit of questions of cultural meaning," with a specific focus on seeing how the workings of "science in public"—an approach that is "underresearched and undertheorized"—can lead us to a better understanding of how science actually operates at any given time and place in its history (Pandora, 347). The Freeman case provides a view into both some key aspects of the everyday conduct of political activity and some of the broader cultural stakes of medical science. Bringing into high relief the interplay between science and politics, it allows us to see patterns that were commonplace for contemporaries, but have eluded more modern eyes because our categories of analysis lead us to prioritize different dynamics, to create different objects of historical knowledge.

The 1846 murder trial of William Freeman raised this multilayered interplay between politics and science into high relief. By itself, it was not an exclusive turning point. Many of these mutually reinforcing trends and ideological tensions had been developing for a decade or more. And as influential as the Freeman case proved to be (it is still a touchstone for both legal precedent and popular culture today), the contentions that it raised for resolution carried on after its conclusion. That said, it was a landmark case of the era. Furthermore, by bringing these widely influential but sometimes tacit views on science and politics before the bar, participants in the conflict were compelled to give clear articulations of their social views and epistemological assumptions. Beyond this, the courtroom setting brought together different strands of political and scientific dialogue that often typically took place in disparate settings. By making explicit the points of contact between them—the warp and woof of these cultural conversations—we get a better picture of the dimensions of the fabric of this struggle over knowledge and moral authority.

A Setting for Strife

The path that led William Freeman to attack the home of John Van Nest and his family on March 12, 1846 was both strange and all too typical. Born in Auburn, New York, in 1824, Freeman grew up in the period immediately after New York abolished slavery in 1827. His father died when he was very young, and with a large family to take care of, his mother put Freeman out to work in a variety of white families starting when he was eight. Although he lived under the pervasive racial exclusion and discrimination common to the era, it was at age sixteen that his life took a dramatic turn for the worse. Freeman was falsely accused of horse theft (the key witness was the actual thief), and sentenced to five years of hard labor at the New York State Penitentiary at Auburn. Although designed as a reform institution, based on the increasingly popular environmentalist understanding of human nature, the prison at Auburn had precisely the opposite effect. Beaten repeatedly over the head with a wooden plank by a prison guard, something changed in Freeman, and he was thereafter widely known as "Crazy Bill."

Upon his release from prison, Freeman managed to carry on with his life for six months, living with acquaintances for a month or two at a time, and piecing together a variety of short-term jobs. But he remained fixated on the injustice of having lost five years of his life, without any compensation. He ultimately determined that he should "do something to get my pay" (Hall, 248). He later talked of how he needed to "kill around" to get his pay (Hall, 348). Freeman's exact target, however, remained obscure, even after the attack. On the one hand, he planned for the event several days in advance, fashioning both a dagger and a bladed cudgel at local artisans' workshops. On the other hand, on the night of the attack he headed south of Auburn by several miles, trying a couple of houses (which had locked doors or insufficient light to do his "work") before settling on the Van Nest house. Furthermore, during the assault itself, Freeman left a significant amount of cash on the dining room table, and thus his understanding of "pay" seemed to have shifted by that point. In the process of killing four members of the Van Nest household and wounding another, Freeman himself received a deep wound on his right wrist, preventing him from continuing "my work" (Hall, 272). He took a horse from the barn, rode through the night to a small town north of Syracuse, and was apprehended the next day by the posse sent out to find him. Brought back to the Cayuga County Jail in downtown Auburn, Freeman became the source of endless fascination and speculation, as people tried to make sense of what could bring him to commit such atrocities.

The Freeman case took on meanings that it likely would not even have had a generation before, since the political and scientific setting in America in 1846 was significantly different even from a quarter century earlier. Along with all of the socio-economic changes typically cited by historians—the "transportation revolution" and the "market revolution" being chief among these—the nation experienced a shift away from the lack of political parties in the early 1820s (the so-called "Era of Good Feelings") to the bitter partisanship of the 1830s and after. Less well known, except by specialists in the field, is the shift from general support for organized medicine to critique of organized medicine. Even less studied—but most relevant for this essay—was the shift from broad cultural agreement on the need for social reforms (and on the scientific foundations for these) to mounting critiques of both such reforms and their conceptual foundations. By 1846, however, the tensions entailed by this shift were sufficient to induce two national-level political leaders to engage in an otherwise obscure criminal case as a means of forwarding their respective positions on these matters.

Both William Seward and John Van Buren were deeply ensconced in the political transformations of the era. Not long after Seward moved to Auburn to begin his legal career, he attached himself to the Anti-Masonic Party formed in New York in response to the actions and exclusionary practices of the Masons. This party eventually became one component of the Whig Party in the mid-1830s, and Seward rose to prominence along with it. Elected as Governor in 1838 and again in 1840, Seward would eventually be a contender for the Presidency, and serve as Abraham Lincoln's Secretary of State. At the time of the Freeman affair, then,

Seward was a figure of national prominence, but in between official positions, and practicing law from his Auburn office.

John Van Buren was similarly involved in politics, although prior to the 1840s this was more often as political advisor to his father Martin. A founder of the famous "Albany Regency" political faction in New York, Martin Van Buren rose to the Governor's mansion in 1828, and voters returned him there in 1830. The elder Van Buren assisted Andrew Jackson in his bid for the U.S. Presidency, and was rewarded with an ambassadorship to England (accompanied by his son John) and later with the Vice Presidency. A stalwart of the Democratic Party that centered on Jackson, Martin Van Buren captured the White House in 1836, although he was deeply disappointed with the sweeping Whig Party victory of 1840. An advisor and companion to his father through much of the 1830s, John Van Buren took more of a place on the political center stage with his ascension to the office of Attorney General of New York in 1844, and with his organizational and campaigning role for the Free Soil Party (for which his father Martin was a presidential candidate) in the later 1840s.

What did the Freeman case represent for each of these prominent politicians? For William Seward, the Freeman case was, on the one hand, an interruption and a burden, but on the other hand, a chance to pursue some deeply held values. After two terms as the Governor of New York (1838-1842), the social expenses of which left him deeply in debt, Seward had returned to his adopted hometown of Auburn to resume his law practice. When Freeman attacked the Van Nest family in March of 1846, Seward was midway through a thousand-mile circuit trip, pursuing lucrative patent law cases across the Mississippi River valley. When he cut short his trip to return to New York to pursue what would likely be a lengthy (and unremunerated) criminal trial, then, Seward undercut his most immediate personal interests.

At the same time, Seward took on Freeman's cause in order to forward his (and his fellow Whigs') reform agenda. When Seward and the Whigs were in political power in New York (beginning in the late 1830s), they led the way in establishing new schools, prisons, insane asylums, and a variety of other reform institutions. All of these were conceptually founded upon an environmentalist understanding of human nature, wherein a person's mind was significantly shaped by their body, which in turn was largely a product of its physical and cultural environment. (Howe, 1997) Reformers thus leaned heavily on the authority of science, even though there was little explicit explanation as to how this formative process worked.

Until the mid-1830s, this advocacy of social reform and its scientific foundations in environmentalism were widely shared. Reformers and their ubiquitous organizations "comprised a virtual proxy nation that counteracted the centrifugal forces of the market revolution and partisan politics" (Brooke, 20). At least two influences changed this, however. The first was the development of political parties themselves, slowly coalescing in the late 1820s and early 1830s. In general, the Whigs were keen to invest in both physical "internal improvements" (e.g. Erie Canal, Maysville Road) and social reforms aimed at bettering the human condition. Government intervention would, by one means or another, help all citizens. Many Whigs emerged from a political tradition that "held that government at all levels of the federal system should be used to positively elevate people economically, socially, and morally through the internal development of the nation's civic institutions and economic infrastructures" (Holt, 951-952).

Democrats tended to oppose anything beyond a minimalist government, and often believed that the fruits of such spending would help only select groups within American society. The Democrats questioned all kinds of government spending, and they made projects such as internal improvements and reform institutions their particular target (Ashworth, 42-43). In his second annual message to Congress in 1830, Jackson even questioned the necessity of improvements to "ports of entry," such as "[l]ight houses, beacons, buoys, public piers, and the removal of sand bars" (Brown, Timothy & White, John 96). He argued that such improvements were often "local" in character, and that those people closer to them (as with those closer to inland improvements like the Erie Canal) would benefit disproportionately to those at a greater remove. Jackson also famously vetoed fiscal appropriations for the interstate Maysville road, put forward as a means to promote interstate commerce.

In both rhetoric and voting action, then, Whig politicians were far more likely to support internal improvements. In Pennsylvania during the early 1840s, for example, they were twice as likely to vote this way. When it came to funding humanitarian or social reform endeavors (e.g. schools, poor houses, facilities for the deaf), Whigs support was sometimes five times as high as Democrats (Egnal 65-68). As frequently happens in politics, anything that one's opponents propose becomes anathema.

What really added the specific vehemence against the use of scientific arguments to support humanitarian endeavors, however, was the extension of these to the issue of slavery. Prior to the application of an environmentally-based humanitarianism to the issue of slavery, there was very little blowback. Indeed, even though the foundational scientific model of environmentalism was explicitly invoked in the very heated debates about race since the 1780s (e.g. Samuel Stanhope Smith), no pervasive backlash occurred. Once one contingent of social reformers extended the logic of environmentalism to the most hot-button issue of the era, there was a backlash against the very model of reform.

Beginning in the 1820s, physicians found themselves called upon to explain how the environment influenced the body, which in turn influenced the brain, leading to new and better behaviors (Mullins, 45-60). In this era, far from being a distinct field, medicine was part of the broader scientific discussion, as "science was one of the important influences in shaping the structure of medicine in the nineteenth century" (Bynum, xii). Thus when humanitarianism became controversial, leading it to become an object of political struggle, the science undergirding humanitarianism came into question. Although not wishing to cast aside *all* aspects of science, this assault did lead critics to portray humanitarianism as misguided, at best, and an active assault upon traditional values, at worst.

Expertise or Elitism?

One of the most persistent debates in the Freeman affair concerned the status of testimony put forth as scientific knowledge. Seward's case for the defense relied heavily upon medical expert witnesses brought in from across New York State, all of them holding positions in some of the leading institutions of the day. Modern scientific learning, he argued, should overcome the kind of folk wisdom and popular prejudices that had often guided assessment of insanity, especially in criminal cases. Van Buren and his co-counsel questioned not only the accuracy of particular medical assessments of Freeman, but also the very notion that some people's opinions—no matter what their station—should be given more weight than those of other people. The debate over the neurological condition of Freeman's brain, then, quickly took on shape of the Whig-versus-Democrats political divisions of the day.

The Freeman case provided Seward with a platform to further this position, and he accordingly called in scientific experts from across the state, with whom he promised to give the court "a proof of the difference between this real science and the empiricism upon which the Counsel for the People rely" (Hall, 384). Seward summoned five medical specialists from across the state, along with four local physicians, to take the stand in Freeman's defense. Pride of place was given to Dr. Amariah Brigham, the medical superintendent of the recently opened New York State Asylum at Utica. Brigham came with an impressive background of both training and experience. In addition to his formal medical studies, he had also spent time in the major hospitals of Britain and France. Further, prior to taking on the New York asylum directorship in 1842, he had worked in Connecticut's insane asylum at Hartford, and thus had more than a decade of direct daily experience in working with cases of insanity. Brigham had also published a book on neurophysiology, *An Inquiry into Diseases and Functions of the Brain, Spinal Cord, and Nerves* (1840), as well as one on the social and environment causes of mental derangement, *Remarks on the Influence of Mental Cultivation and Mental Excitement upon Health* (1833). At both the preliminary hearing on mental competency and the actual trial on murder, Seward placed Brigham as the final witness, so that the doctor could not only give his own testimony, but also analyze the testimony that preceded his own.

The other physicians whom Seward called to the stand were also notable not merely for their prominence in the medical profession, but because they were situated as the most "scientific." This stature was based on several factors. It came in part from their having attended medical school in an era when a large portion of physicians still received their training through a single apprenticeship. They were also medical professors themselves. And they joined—and were leaders in—medical societies. This allowed them to present their own research (and doing research was a rarity, indicating that they were indeed involved in medicine as *science*), and to stay abreast of developments in medical science by hearing the research presentations of others, and reading about research developments from afar (in their journals).

By the 1840s it was increasingly common to call physicians as witnesses to testify in court. While cases of criminal insanity were one example of this, lawyers brought them in for cases of suspected poisoning, contested wills, and a whole range of other matters. As James Mohr has underscored, "the early champions of medical jurisprudence constituted something of a professional elite" (Mohr, 77). The authority to which they laid claim on the basis of both their experience and their scientific models often were given credence. Contemporaries were careful to distinguish between experience and scientific knowledge, and many lay witnesses who took the stand in the Freeman case demonstrated that they understood the limitations of the insights into mental illness. When Van Buren cross-examined defense witness Ira Curtis as to whether he was "acquainted with the subject of insanity," Curtis replied that, although he had "seen many cases of it," he had to acknowledge that, "I don't know any thing about it scientifically" (Hall, 39).

The prosecution took a very different view of the testimony of such witnesses. In opening the case for the prosecution, Van Buren's co-counsel Luman Sherwood (District Attorney for Cayuga County) set the tone for the next month of legal proceedings. A Democratic political appointee, Sherwood wasted no time in challenging the learning of expert medical witnesses for the defense, sometimes crossing over into mockery. In his opening address at the murder trial, Sherwood told the jury that they were "to receive lessons upon the subject of insanity," delivered from "Witnesses, learned witnesses, in [sic] behalf of this prisoner." At their hands the jurors would "be instructed in the various classifications and divisions that science ...has made of mania" (Hall, 178). Sherwood promised the jury that, "before the close of this trial," they would see "how men of learning and of science may be deceived by converting natural imbecility and taciturnity into strong 'symptoms' of mental derangement." Indeed, Sherwood continued, reflecting back on the testimony given in the preliminary hearing by this same witnesses, that "the disquisitions of learned doctors of medicine and of mind, upon the symptoms of insanity," he was "almost led to exclaim, in the language of Festus to Paul, 'Thou art beside thyself; much learning doth make thee mad'" (Hall, 174). So, the prosecution pulled in biblical images to counter the claims of science.

The prosecution's invocation of the Bible was no accident. In making their case for the prosecution, Van Buren and Sherwood frequently tapped into the concerns about elitism that they believed would play well with jury of Cayuga County residents. There were at least two veins of this American debate on elitism, and both made an appearance in the trial. One of the most pervasive contributors to a tendency to the sentiments of anti-elitism in the country emerged from the sphere of religion. In an era of revivals and religious populism, many communities explicitly expressed skepticism about the spiritual insights that could come from a formally educated ministry (Hatch, 18). Another relevant trend was the removal of qualifications to vote, so that by the 1830s nearly every adult white male had the franchise, a substantial shift from two decades earlier. This led many to conclude that any distinctions one might make between two groups' capacity for moral insight were specious. Both of these developments played a role in the Freeman trial.

Such attitudes were not merely cultural biases, nor even limited to electoral contests (i.e. mud-slinging to gain votes). They could have profound implications for the status of knowledge. In the first half of the nineteenth century, learning itself was often viewed with suspicion. Many communities did not want a minister who had attended seminary, or even had any advanced education. In a few cases, even literacy was looked upon with suspicion. Such views were not universal, but the dramatic success of religious revivalism—with its embrace of emotional immediacy and rejection of dry, learned theology—underscores the prevalence of such attitudes.

In making such a case, Van Buren and Sherwood were tapping into a political ideology articulated at the highest levels of the Democratic Party. Even before the mid-1830s rift over humanitarianism, Andrew Jackson had articulated a profound disdain for expertise. During the opening months of his presidency, this guided Jackson's choice of appointees to political offices. By the time of his first annual address to the U.S. Congress, President Jackson found it necessary to defend his approach to appointments to federal offices. Making a claim for the intrinsic equality of all men, Jackson argued that, "no one man has any more intrinsic right to official station than another." In practice, however, this argument was often extended to mean that there were no particular qualifications necessary for any government position. Jackson himself did not believe that the positions in the various arms of government (e.g. judiciary, military, treasury) required any special expertise. In 1829, in his first annual address to the U.S. Congress, he explained that, "The duties of all public officers are . . . so plain and simple that men of intelligence may readily qualify themselves for their performance" (Brown, Timothy & White, John, 48).

This anti-elitism, gaining national currency during the Jackson administration, was particularly strong in New York, a major seedbed of such views. Martin Van Buren had helped to set the tone during the 1820s, when he made no secret of political offices and bureaucratic positions being solely a matter of patronage. In 1830, Senator William Marcy of New York (himself a member of Van Buren's "Albany Regency") had come most explicitly to Jackson's defense on the question of political appointments. New York Democrats having long embraced this type of appointment, Marcy was not afraid to call it what it was: the spoils of war. In a speech before U.S. Senate, Marcy argued that he "saw nothing wrong in the rule, that to the victor belong the spoils of the enemy" (Spencer, 182). Van Buren and Sherwood's assessment of the value of the defense medical expert witnesses, then, was well in keeping with their party's ideological stance.

By attacking the status of both the physicians and the medical science that they proffered, the Democratic Party-aligned prosecutors furthered a political position that was well established by the 1840s. Van Buren disparaged Brigham's expertise, observing that, "while he is innocently dealing with doctrines that may convulse Society to its centre, he reminds me more of a child playing with Lightning, than a Scholar and Philanthropist bringing the lights of Science to aid the investigation and establishment of truth" (Hall, 443). Van Buren charged that "misguided philanthropy" drove Seward and other reformers, leading them to deploy science in service of goals with dangerous implications.

The defense did not sit idly by in the face of this characterization of learning and expertise. Seward's co-counsel David Wright pointed out to the jury that the expert witnesses had traveled far to be there, without compensation, yet when examined by Sherwood "they must be subject to his ridicule." Why? Only because "their learning, their experience, and ...their wisdom, does not happen to coincide with the opinion of the learned counsel, in regard to the condition of the mind of this particular subject" (Hall, 224). But it was precisely this medical learning and experience—in a "science, the most difficult in the range of human studies"—rather than a lawyer's "opinion," that Wright urged the jury to accept.

Seward took the argument even further. Not content to simply have the expert medical witnesses make their particular arguments about Freeman's specific state of mind, Seward himself invoked the language of science. Wishing to resonate with the jury members on a less abstract level, he posed the question, "And now what is Insanity? Many learned men have defined it for us, but I prefer to convey my idea of it in the simplest manner." Answering his own question, he stated simply that insanity was "disease of the body, and I doubt not of the brain." Following immediately upon this, however, Seward made his larger point, expressing surprise that, "The world is astonished to find it so." Implicitly responding to the prosecution's arguments that modern scientific explanations of insanity were too new to be relied upon, Seward worked to normalize the reliance on scientific naturalism in determining moral accountability. He not only brought the medical experts to the stand (reinforcing their testimony in his closing arguments), but he made appeals to scientific development himself. While some might find it "strange that the discovery should have been made so late," he reminded jurors of the railroads just becoming common in the region. It now seemed obvious that it was easier to move people and cargo "upon two smooth rails on a level surface, than over the rugged ground," yet it has "taken almost six thousand years to learn that" (Hall, 380). By comparing ideas about insanity to ideas about commerce and transportation, Seward attempted to make more accessible the notion of medical concepts as discoveries resulting from the unfolding of truths of the natural world.

Seward worked to accomplish multiple goals with these kinds of arguments. In order to distance his immediate case for Freeman from the broader cultural and political contest for medical authority, Seward tried to convey to the jury his view that recent concepts of insanity were not opposed to either common sense or sound moral judgments. He also carefully framed his point in biblical language, citing "six thousand years" as the age of the earth (and humanity), when scientific understandings of geology of the time generally cited a much older origin. He thus hoped to mitigate the impact of the anti-elitism and suspicion of formal learning that emerged from the revivalism of the preceding generation. Seward also worked to get religion on his side when he exhorted the jury, "If you are bent on rejecting the testimony of those who know, by experience and by science, the deep afflictions of the Prisoner, beware how you misinterpret the handwriting of the almighty" (Hall, 419).

Just as Seward simultaneously asserted naturalist principles and appealed to popular religious sentiments, so too did he aim to strike a balance between invoking scientific authority and refraining from trotting out a lot of technical details. With Amariah Brigham in the courtroom, Seward had at his disposal not only one of the country's most experienced physicians, but also one of America's leading authors on neurophysiology. Indeed, Brigham figures into some of the modern histories of the development of neurophysiology (Clarke, Edwin., Jacyna,L.S., 9-13). Many of the defense witnesses were students of brain science, but Brigham was an active contributor. Nonetheless, the defense never called upon him to present the scientific details of his expertise. In an era when expertise itself could be seen as a liability, Seward had walked a fine line as he pushed forward his case for a naturalist understanding of the human condition.

In the decades before the Civil War, Americans wrestled with two fundamental political questions: what was the best interest of society, and who was best positioned to pronounce on this? Perennial issues for any polity, these questions grew especially prominent in the U.S. due to the egalitarian rhetoric built into the political culture. In the realm of science, two corollary questions were: what was genuine science, and who had the right to pronounce on what is a scientific finding? As applications of medical science came to have larger and more controversial implications for society, science found itself subject the strife and pitched infighting already so common to elections and legislative battles. This meant that people generally (versus learned experts) increasingly weighed in on questions of scientific truth.

Medical Monopoly, Moral Monopoly

In both the Freeman trial and American culture broadly, one issue related to debates about elitism was a deep-seated concern about monopoly. In standard narratives of American history, the concerns about monopoly are usually cast as residing principally in the political sphere. Most famously, this took the form of Andrew Jackson's pitched battle to destroy the Second Bank of the United States, charging that it allowed access to funds only to a select clique of wealthy elites. This was, however, only the most high profile instantiation of American fears of monopoly. In a wide variety of spheres, there was a concern about exclusion from access to the means for social and economic prosperity.

As Van Buren made his closing argument to the jury, he emphasized both the inappropriateness of medical experts declaring upon moral matters, and the hazards that such a situation might entail. "Did the detection of offenders or the prevention of crime depend upon medical skill," he argued, "our police should be composed of Physicians and Nurses" (Hall, 428). To allow such persons too much influence in the courts, they would usurp the role of everyday citizens, as "the Professors should sit upon the bench and fill the jury box." Even Van Buren's own role would be usurped, since the Governor would have "sent the Surgeon General instead of the Attorney General to assist at this trial." Fortunately, "the law allows no such absurdities" (Hall, 428). The democratic foundations of American government would be overturned, Van Buren argued, if Seward's medical expert witnesses were allowed to pronounce upon Freeman's moral accountability, and thus his guilt or innocence.

The tensions in the Freeman affair concerning a medical—and thus moral—monopoly emerged from the political tensions of the prior decade. Perhaps the signature example of this was the assault by the early Democratic Party on the Second Bank of the United States. Although the "Bank War" has typically stood in as exemplary of the anti-monopoly sentiment of Jacksonian Democrats, it is far from being a unique case. Contemporaries complained about all kinds of monopolies. We can see this in the struggle over whether storeowners and street vendors in New York City would need a license to ply their trades. The newspaper the *Working Man's Advocate* spoke in 1834 on behalf of the working classes, saying: "Almost every article necessary to their existence… is encumbered with an indirect tax, levied through the agency of this license humbug." (quoted in Hugins, 163) Another newspaper, *The Democrat*, in 1836 framed it thus: "we cannot pass the bounds of the city without paying tribute to monopoly; our bread, our meat, our vegetables, our fuel, all pay tribute to monopolists" (quoted in Hugins, 163). The attack against the license rules was led by a political faction known as the Workingmen, an independent group that would eventually be absorbed into the Democratic Party. Their concerns went beyond the immediate concerns of the pocketbook, to encompass broader ideological points. As one scholar of the protests has noted, "It was not only as consumers but as producers and entrepreneurs that the Workingmen reacted to this regulation," since licensing could promote favoritism (in areas of endeavor licenses were limited), and a restriction on individual liberty to practice a trade (since there could be fines or even imprisonment for violations) (Hugins, 163).

The Workingmen also challenged what they termed "the medical monopoly." (Hugins, 166-171). In what might seem like an odd alliance, they made common cause with medical practitioners who were adherents of Thompsonianism, an alternative medical model that was generally despised by "regular" (allopathic) physicians. Up until the 1830s, state medical societies had enjoyed considerable success in convincing legislature to pass laws requiring either an M.D. (from a medical school) or an examination by the medical society in order for someone to practice medicine. During the 1830s and 40s, however, the growing antimonopoly sentiment led to a reversal of much of this legislation. As one historian of this sweeping change has noted, "Within the space of little more than a decade, however, nearly every state had repealed its penalties on unlicensed practitioners" (Kett, 13). It was not just that the issue of medical licenses as part of a monopoly played into a more pervasive concern about licenses and monopoly. Proponents of specific views of medical science forged alliances with the particular political parties that had compatible ideological views. So it was that proponents of alternative medicine (the Thompsonians, in New York) not only embraced aspects of Democratic political ideology, but also allied with the Loco Foco Democrats. Together they took up the medical license cause directly, and by 1844, the anti-license movement in New York met with success.

While both the political and medical scientific landscapes were too complex for there to be absolute polarization, pushing one group of medical practitioners always into the arms of a particular political party, the elective affinities existing amongst them did produce some enduring links. Whigs saw qualifications as preventing fraud and ensuring quality, and so attracted those physicians wanting a more professionalized and regularized medical community. Democrats saw the push to legislate such qualifications as oppression, and so tended to attract those doctors approaching medical science from alternative perspectives. As in New York, so too in neighboring states did the anti-license movement become a widespread political battle that typically split along Democrat-Whig lines (Appel TA, 178-181).

One motivation for these expert physicians to travel hundreds of miles, for no compensation, in this case was the chance to bolster the status of "regular" medicine in general, and the status of their home institutions in particular. Just the year before the trial, for example, Dr. Charles Coventry, one of the expert witnesses for the defense in the Freeman trial, reflected back upon these events, arguing that, "when restrictive laws are really efficient and enforced, they protect the community against inexperience and its consequences," but that "popular sentiment is opposed to them." As a result, "the law is either so drawn as to be inefficient, or is, in nine out of ten cases, openly violated with impunity, whilst its existence is such as to get up a feeling of hostility to the regular profession" (Coventry, 160).

The political dimensions of the Freeman case ran deeper than the divisive issue of medical licensure. Certainly the recent repeal of New York's license requirement was a prominent memory for both the more professionalizing "scientific" physicians and for their opponents. And since the legislative battle split along political party lines, certainly it was a point of partisan contention for both Democratic and Whig leaders. More profoundly still, however, the Freeman case raised the core issue of what role was the government to play in promoting the welfare (be it economic, educational, or medical) of citizens. Should the social reform endeavors championed by the Whigs—and the environmentalist scientific understanding of human nature at their foundation—be allowed to carry forward? And, as one recent scholar has demonstrated, these competing visions of the good society were fueled by competing understandings of freedom. Hence the struggles over medical licensure carried a much larger dimension, since "the American commitment to freedom of therapeutic choice was based on notions of not only bodily freedom, but also economic freedom, freedom of conscience, and freedom of inquiry" (Grossman, LA, 76-77). The stakes in the Freeman trial ran parallel to these.

Given all of this, there were at least two important questions raised by this ideological tension. First, could professional expertise—however well founded—undermine the foundation of democratic government? The role of medical expert witnesses put this question into high relief. The chief fear was a monopoly on moral judgment, carried out in the name of medical science. Even if the science was accurate, this could lead to a transfer of the moral role of the jury into the hands of physicians, effecting a shift both in the locus of social and cultural power, and potentially towards a mechanization of morality. Secondly, was the status/authority claimed by expert opinion merited? Did they actually deliver the goods? If not, then there was little reason to accord it great weight. And it was not entirely clear that the scientific understanding of the brain and its disorders was anywhere near firm enough to countenance such a shift in the foundations of moral judgment. Furthermore, far less was there reason to exclude all other insights on the basis of such claims to definitive authority.

The Spoils of Science

The position of medical science in the 1840s allowed for multiple appropriations. Indeed, for some, it was possible to utilize aspects of the authority of scientific opinion even as they called such views into question. If Van Buren and Sherwood both mocked the learning of the defense expert witnesses and questioned the validity of the "speculative theories," they concluded that they could not simply eschew medical testimony altogether. Playing to regionalism and the anti-expert bias, they brought eight local physicians to the stand. More importantly, they also felt compelled to have some professionally established and institutionally certified physician to counterbalance Brigham and the four medical school professors testifying for the defense.

So it was that, as much as the prosecution argued that the medicalization of morality threatened to undermine the democratic nature of American law and government, they themselves ultimately ended up deploying scientific evidence, arguments, and witnesses. At the start of the trial, they informed the jury that, "We will prove by medical men... [that there was] nothing that indicates insanity in any of its forms. And upon such proof, aided by science, and by sense, we shall rely for a conviction of the prisoner" (Hall, 182). In part this approach was purely practical. To have allowed the defense to lay out an extensive scientific case for the non-responsibility of Freeman, without offering a competing understanding of the science at stake, would have been an enormous gamble. Furthermore, the value and cultural power of science was too prevalent by the 1840s to ignore.

The principal expert medical witness for the prosecution was Thomas Spencer, a professor at the nearby Geneva Medical College. While he came with the kinds of core credentials that the prosecution sought, it quickly became clear that Spencer was in many ways a problematic witness. Although a physician, and a medical professor, Spencer's medical research focused on cholera. Most importantly, he had almost no prior experience with insanity. When cross-examined, it became clear that he had little experience in broader medical issues. And while he had toured hospitals during a trip to Europe—an increasingly common stamp for professionalizing physicians—under cross-examination he could not remember what he had learned in Europe, or even what hospitals he visited to expand his medical training. That said, it was clear that Spencer was very eager to improve his professional standing, and he went to great lengths to establish himself as a credible witness. His testimony was far more detailed than that of the defense witnesses (though sometimes internally contradictory). Spencer, much more so than his colleagues across the aisle, made frequent references and assertions about the certainty of his position, guided by "the truth of science" (Hall, 128). At one point, Spencer makes a conclusion founded upon "what I know of medical science," but then acknowledged that prior to this, he had only ever been called upon to assess a person's (in)sanity "in one case" (Hall, 131).

Ultimately, the trial concluded in the manner that everyone—including Seward—expected that it would. As a black man who had killed four members of a white family in an era marked by systemic racial inequality, there was little chance that Freeman would find a welcome audience from an all-white jury. In July of 1846 he was found guilty, and sentenced to be hanged in September. There were enough procedural errors (including many regarding the court's treatment of expert medical witnesses) in the trial, however, for Seward to successfully appeal to the New York State Supreme Court for a re-trial. By the time Freeman was re-examined in February of 1847, his mental deterioration had progressed so far that even the original judge (a Democratic appointee) acknowledged that he could not in fairness be deemed mentally competent.

Even after the trial was over, the incident continued to be important for both science and politics. Once Freeman had passed away in the summer of 1847, many of the medical expert witnesses flocked back to Auburn to attend the post-mortem examination of his body, particularly his head. Most of the physicians found Freeman's brain to show significant signs of physical degeneration, although there was a vocal minority who differed on this point. Although the autopsy findings had direct implications for Freeman's moral responsibility (or lack thereof), the main concern seemed to be in how his body could advance scientific knowledge. One of the physicians involved observed that, seeing the results of the autopsy, "over prejudice and error, science has gloriously triumphed" (Hall, 506).

In the political sphere, the Freeman affair had a complicated legacy. In some respects, the controversy itself quickly subsided. Both the victims and their attacker were dead, allowing no further possibility for an alternate resolution. Both Democrats and Whigs had other political challenges facing them, and moved on to the next bone of contention between them. In other regards, however, the Freeman case had lasting political resonance. For Seward, the case helped to further his political career, giving him the humanitarian bone fides that allowed him to be the moral voice for the future Republican Party. In retrospectives of his career—both while he still lived and posthumously—many writers lauded Seward's "heroic" stance in the face of "the foaming wrath of the community," securing him a place as a champion of the oppressed. Blanchard Fosgate, one of the local physicians involved in the trial, wrote: "How much the cause of justice and philosophy is indebted to the unwearied perseverance of the eminent advocate who withstood the tide of popular indignation in conducting the prisoner's defence, is left for other hands to register; but true it is, that over prejudice and error, science has gloriously triumphed" (Hall, 506). If John Van Buren and the Democrats secured the legal victory, William Seward and the Whigs arguably won a moral victory.

More broadly, however, we find in this controversy evidence of the extent to which the meaning of science was very much subject to political debate. As Lorraine Daston and Fernando Vidal have observed, "why and where Nature's authority is called on to buttress or subvert human norms requires a careful sorting out of the meaning s of both nature and authority in specific contexts" (Daston and Vidal, 4). There is no reliable abstract generalization that one can make about the linear direct of the development of scientific knowledge. Rather, it is in the everyday practices that one finds such meanings forged. Often, though, everyday practices remain opaque to historians precisely because of their quotidian character. A controversial case such as this one, calling out so much political and scientific firepower to central New York, can help to illuminate these practices and thoughts. And, turning to Daston and Vidal again, "the very existence of such controversies helped call into existence both nature as an allegedly neutral judge and experts as allegedly disinterested interpreters of nature's verdicts" (Daston and Vidal, 7).

Bibliography

Appel T. A. 2010. The Thomsonian Movement, the Regular Profession, and the State in Antebellum Connecticut: A Case Study of the Repeal of Early Medical Licensing Laws." *Journal of the History of Medicine and Allied Sciences* 65.2, pp. 153-86.

Ashworth, John. 1987. *"Agrarians" and "Aristocrats": Party Political Ideology in the United States, 1837-1846.* Cambridge [Cambridgeshire], New York: Cambridge University Press.

Brooke, John L. 2009. "Cultures of Nationalism, Movements of Reform, and the Composite-Federal Polity: From Revolutionary Settlement to Antebellum Crisis." *Journal of the Early Republic* 29.1 (2009), pp. 1-33.

Brown, Timothy & White, John. 1837. *Messages of Gen. Andrew Jackson: : With a Short Sketch of His Life.*

Bynum, W. F. 1994. *Science and the Practice of Medicine in the Nineteenth Century*. Cambridge [England]: New York: Cambridge University Press.

Clarke, Edwin., Jacyna, L. S. 1987. *Nineteenth-Century Origins of Neuroscientific Concepts*. Berkeley: University of California Press.

Coventry, Charles B. 1845. "History of Medical Legislation in the State of New York ." *New York Journal of Medicine* 4, pp. 151-61.

Daston, Lorraine, and Fernando Vidal. 2004. *The Moral Authority of Nature*. Chicago: University of Chicago Press.

Egnal, Marc. 2009. *Clash of Extremes: The Economic Origins of the Civil War*. New York: Hill and Wang.

Golan, Tal. 2004. *Laws of Men and Laws of Nature the History of Scientific Expert Testimony in England and America*. Cambridge, Mass.: Harvard University Press.

Grossman LA. 2013. "The Origins of American Health Libertarianism." *Yale Journal of Health Policy, Law, and Ethics* 13.1, pp. 76-134.

Hall, Benjamin F. 1848. *The Trial of William Freeman for the Murder of John G. Van Nest : Including the Evidence and the Arguments of Counsel, with the Decision of the Supreme Court Granting a New Trial, and an Account of the Death of the Prisoner, and of the Post-Mortem Examination of His Body by Amariah Brigham, M.D., and Others*. Auburn: Derby, Miller & Co.

Hatch, Nathan O. 1989. *The Democratization of American Christianity*. New Haven: Yale University Press.

Holt, Michael F. 1999. *The Rise and Fall of the American Whig Party : Jacksonian Politics and the Onset of the Civil War*. New York: Oxford University Press.

Howe, Daniel Walker. 1997. *Making the American Self : Jonathan Edwards to Abraham Lincoln*. Cambridge, Mass.: Harvard University Press.

Hugins, Walter Edward. 1960. *Jacksonian Democracy and the Working Class, a Study of the New York Workingmen's Movement, 1829-1837*. Stanford: Stanford University Press.

Kett, Joseph F. 1968. *The Formation of the American Medical Profession; the Role of Institutions, 1780-1860*. New Haven: Yale University Press.

Mohr, James C. 1993. *Doctors and the Law : Medical Jurisprudence in Nineteenth-Century America*. New York: Oxford University Press.

Mullins, Jeffrey A. 1995. "Duties to Science, Duties to God : Medical Theory, Physiology, and the Discourse on Morality in Nineteenth-Century America." *Transactions and Studies of the College of Physicians of Philadelphia* 17, pp. 45-60.

Novak, William J. 1996. *The People's Welfare : Law and Regulation in Nineteenth-Century America*. Chapel Hill, N.C.: University of North Carolina Press.

Pandora, Katherine. 2009. "Popular Science in National and Transnational Perspective : Suggestions from the American Context." *Isis* 100, pp. 346-58.

Spencer, Ivor D. 1941. "William L. Marcy: 'An Educated Northern Democrat'." *New York History* 22.2, pp. 180-90.

'Who Is To Decide Who the Medical Charlatans Are?': Benjamin O. Flower, "Medical Freedom" and Scientific Expertise in the USA in the 1910s

JEAN-LOUIS MARIN-LAMELLET

According to Boston editor and reformer Benjamin Orange Flower (1858-1918) turn-of-the-century America, racked with urban squalor, labor conflicts, political corruption and arrogant plutocrats, threatened to collapse. Flower wanted to use the authority of science to regenerate and modernize the country. Not only did he use Herbert Spencer's evolutionary rhetoric to support his vision of progress, but he also turned to the science of his day, most notably the nascent social sciences, to legitimize new, allegedly "cranky," reform ideas such as Populism, woman suffrage, direct legislation, and public ownership of utilities, to name but a few examples. Flower used his reform magazines, in particular *The Arena* and *The Twentieth Century magazine,* as forums where new ideas could be discussed and where, out of open debates, truth could be found. As he put it: "*The Arena* gives all sides of every great issue presented by the strongest advocates, as the editor believes that it is only in the crucible of free discussion that we find the gold of truth" ("Praise for the Arena" n. pag.).[1] Between 1910 and 1915 Flower discussed the role and definition of science in a modern society. He focused all his moral fervor on a crusade for "medical freedom." This article explores the controversies over politicizing scientific expertise at a time when the American Common Sense tradition was giving way to a more bureaucratized and technocratic vision of science. In other words, Flower raised a troubling epistemological and political question: who had the intellectual and political authority to decide and impose who the legitimate representatives of "science" were?

Progressives vs. Progressives: the Crusade for "Medical Freedom"

On May 15, 1910, Flower, then editor of the progressive *Twentieth Century Magazine*, set up the National League for Medical Freedom (NLMF) to fight against the "Owen Bill" that Senator Owen (OK, Dem.) drafted in order to establish a federal Department of Health.[2] *Flower saw in his crusade against the bill and what he considered to be its behind-the-scene mastermind, the* American Medical Association *(AMA), the intellectual struggle of* alternative schools of medicine *against the "medical dogmas" that "regular physicians" or "allopaths" wanted to institutionalize.* He claimed that regular doctors would dominate a Department of Health and, with this new authority, become medical monopolies and crush other schools of medicine. Many joined Flower in his campaign: alternative physicians such as leading osteopaths, homeopaths, New Thought practitioners (i.e. advocates of mind-cure), and Christian Scientists (i.e. believers in faith healing); politicians from across the board, Republicans like California Senator John Works and ex-Governor of Massachusetts John Bates, former Populists, like 1892 presidential nominee James B. Weaver (a Democrat in the 1910s) and Senator William V. Allen; as well as many rank-and-file Americans, mostly Westerners according to the *New York Tribune* (*New-York Tribune* 12; *First Report* 6-7; *Congressional Record* 2671).

On the other side of the debate, major figures defended the Owen Bill. President Roosevelt, in his message to Congress in 1908, had called for such a reform. Taft followed suit in 1909 and, in 1910, regretted the "unwarranted counteragitation" that allegedly held back progress. Many others supported the bill: Republicans and Democrats in their respective health planks; insurance companies; unions; farmers' organizations; the AMA, and the Committee of One Hundred on National Health (Fisher "National Health" 514 and "Defines Opposition" 8; "Organization of Ill Health" 421). The Committee had been set up by the American Association for the Advancement of Science. Its president was economist and Yale professor of political economy (and later eugenicist) Irving Fisher, and it counted as members reformers like social worker and public intellectual Jane Adams, pastor and magazine editor Lyman Abbott and president of the United Mine Workers John Mitchell (Knopf, 374).

All these Progressives *applauded this move toward the centralization of several health bureaus into a more efficient Department of Health.*[3] *Science would be used to* "improve public health" by educating the public about diseases, controlling the misbranding of food and drugs and organizing health inspections of the food industries as well as quarantine regulations in case of epidemics (Fisher "Defines Opposition" 8). *As Fisher claimed,* "the world is clamoring for progressive regulation": *the policy had proven beneficial in France and Germany, as shown by the successful examples of Pasteur and Koch's government-assisted fight against diseases (Fisher Memorial, 71, 75;* Knopf, 377).

In reforming circles, Flower had "the reputation of a fighter for everything involving the spiritual, social and physical progress of humanity." It was therefore "inexplicable" to "his admirers" how such a fearless reform editor could "lead a movement opposed to the improvement of the health of the nation" (Knopf, 376). Whatever the dismay expressed by "his admirers," Flower and his supporters did consider their crusade as reformist–it crowned *more than twenty years* of struggle *for a fairer society a*nd, accordingly, *concluded his* memoir of the progressive movement (*Progressive Men,* 298-316). Flower had actually been standing up for medical freedom since the end of the 1880s when he had fought against restrictive legislation in Massachusetts (*Progressive Men,* 314). In different *Arena* editorials in the 1890s, Flower had denounced the "class tyranny" that imposed AMA-sanctioned medicine on citizens. In States like Iowa, medically restrictive laws outlawed alternative cures. For instance, Mrs. Lottie Post, an unlicensed Christian Science physician was fined for curing a woman after an unsuccessful treatment by a "leading physician," which prompted Flower to claim that it had "become criminal in many States of the Union to cure the sick" ("Object Lesson", 507-10; "Class Interests", 510-11; "Herbert Spencer", 511-12). His 1894 editorial, "The Menace of Medical Monopoly," first reproduced in 1898, was then reprinted in *The Twentieth Century Magazine* and used as the textbook for medical freedom in the 1910s.

Flower's struggle at the state level in the 1890s became a nationwide popular revolt when the AMA, with the Owen Bill, threatened to impose medical laws at the federal level. In the 1910s, Flower also included in his fight a crusade against what he claimed were the dangers of compulsory AMA-sponsored medical inspections of school children and vaccination campaigns ("Compulsory Medical Inspection", 151-60; "Rising Revolt", 5; "Bird's-Eye View", 117). As the campaign gathered momentum in 1910, each issue of the *Twentieth Century Magazine* featured articles about "the battle for medical freedom." In October 1911, Flower retired as editor of the magazine to devote all his time to the work of the NLMF. On September 17, 1912, on top of his position as president of the NLMF, he was also elected editor of *Medical Freedom*, the official journal of the League (*Medical Freedom,* Oct. 1912 2). A heated debate raged on until

1916, when the final attempt to pass a new version of bill was defeated and the NLMF finally disbanded ("Closing of the League", 2).

Progressives vs. Progressives: Conflicting Health Discourses

When Flower organized the League in May 1910, he inaugurated the first modern lobbying effort on a large scale by using a new political tool: advertising. The NLMF had neither time nor money to print and mail pamphlets to politicians, intellectuals and doctors. They opted instead for an "immense, simultaneous newspaper campaign from coast to coast". According to *Printers' Ink*, the first national trade magazine for advertising, the campaign demonstrated that the age of free coal for tenement dwellers in winter and free beers at picnics was over. Advertising broke with such 19th century practices, which were reminiscent of corrupt machine politics—it provided a modern, efficient way to influence voters, as it could reach thousands of people fast and without any intermediaries. In little more than two weeks, this "bold display campaign" had spread from New York westward and reached more than 15 cities. As a result, 100,000 petition signers flooded Congress with indignant letters. NLMF offices were reportedly overwhelmed with letters of support, many with money even though the ads did not mention any fundraising. During the Senate hearing on the Owen bill on May 19, Senators Martin and Depew claimed that they had received more than 1,000 telegrams from alarmed constituents—something unheard of in their careers. The bill was killed instantly. *Printers' Ink* concluded: "in advertising lies inestimable power along national political lines" (47-49).

Not only is Flower's advertising campaign interesting in terms of innovative advertising-based mobilization strategies, but it also lays bare Flower's framing of the question in progressive terms. One of these ads read: "Do You Want the 'Doctors' Trust' to be able to Force its Opinions on You?"[4] He denounced the "Doctors' Trust," in other words, AMA experts who wanted to enter politics or, to use Flower's words, "political doctors" pushing for "special privileges" ("Spencer", 512; "Restrictive Medical Legislation", 783). Flower also did go into politics with his lobby, but the difference with the AMA for him was that the NLMF strove to influence politicians but did not want to monopolize official positions of power to the detriment of other schools of medicine. *State-sanctioned* medical monopolies would force opinions on the people and lead to "medical censorship" ("Menace of Medical Monopoly", 411). Trusts, according to Flower, were "ostensibly for the protection of the people, in reality for the protection of a monopoly" and "rob[bed] the intelligent of his inherent right to select his own physician and place(d) him at the hands of a protected class" ("Object Lesson", 507-10). Accepting a Department of Health was therefore "contrary to the growing spirit of opposition to trusts, monopolies and privileged interests, which is marked at the present time," i.e. the reforming spirit of the Progressive Era ("Menace of a National Health Bureau", 234).

Flower wanted to protect "the people" against corrupt "private interests" and compared the AMA to a "tyrannical political machine" that united "political grafters with the medical grafters" ("Compulsory Medical Inspection" 159). He claimed the AMA worked hand in hand with serum and vaccine manufacturers ("Position" 7; "An Earnest Word" 26). His opponents used the same kind of rhetoric. AMA physicians and their progressive supporters accused the NLMF of promoting "the sinister commercial purpose" of the patent medicine industry (*Congressional Record,* 2659; Fisher "Defines Opposition" and "Department of Dollars", 330; "Liberty", 9; "Medical Freedom. Editorial" 10; "Fight 'Doctors' Trust"). Who else could finance the extensive newspaper advertising campaign that lashed out at the so-called "Doctors' trust" in a base attempt to oppose "a popular movement" for public health? ("New Combination", 1791; Fisher "Defines Opposition", 8). *The AMA regarded their struggle as the necessary follow-up work to* Samuel Hopkins Adams' muckraking exposé of the health dangers and widespread fraud of the patent medicine industry in 1905 and its legislative outcome the 1906 Pure Food and Drug Act. *Collier's* magazine, which had initiated the exposure, even saw the NLMF as an "octopus"–a master trope in progressive discourse ("Bad Bunch" 10). Did not patent medicine companies advertise in Flower's magazines? Did he not claim that proprietary medicines were "less dangerous" than physicians' drugs? Had he not been president of the R.C. Flower Medicine Company from 1885 to 1899? Richard C. Flower, Benjamin's brother and also a veteran of the fight against "allopathic Czars," was a "notorious quack" and a con artist who got arrested in 1908 ("Liberty", 9). Of course, the NLMF and Flower denied the allegations. They claimed that the articles in *Collier's* had their facts wrong–Flower had been devoting his time to editorial and literary work and he had not had any contact with his brother's company or invested in patent medicine businesses since he started *The Arena* in 1889, *not* 1899. They also pointed out that most NLMF funds came from voluntary subscriptions and that, anyway, it was commonplace for magazines and newspapers to resort to patent medicine advertisements ("Try to stop", 2; *Congressional Record,* 2669-71).

The crusade for medical freedom was not only in tune with the most advanced political ideas according to Flower but also intrinsically American. He legitimized his campaign by referring to American history. For example, he invoked abolitionism, a major historical root (with temperance movements) of turn-of-the-century reform movements and the main metaphor for reformers to express their moral protest after the Civil War (Foner, 438). States with restrictive medical laws, like Iowa and New York, were called "medical slave states" and states like Massachusetts, which had resisted restrictive legislation "medical free states" ("Menace of Medical Monopoly", 415; "Menace of a National Health Bureau", 240). *Medical Freedom*, the magazine, fought for the meaning of American principles and used for its covers all the traditional symbols of American liberty: the Liberty Bell, the Statue of Liberty, the Star-Spangled Banner and presidents like Washington and Lincoln to name but a few examples. Flower's crusade reenacted the Revolutionary War: the NLMF represented the "spirit of 1776" and "patriots" had a duty to defend American principles ("Fight", Sept. 1910 368; "Appeal", 3). America was supposed to the land of the free and the home of progress. Its genius was in danger and it demanded yet another popular revolt against what he saw as a new form of slavery and the kind of despotism or hierarchies that plagued the Old World and the Dark Ages. According to the NLMF's Declaration of Principles, the purpose of the League was "the maintenance of the rights of the American people against unnecessary, unjust, oppressive, paternal and un-American laws, ostensibly related to the subject of health" (*Medical Freedom,* Sept. 1911 2). The paternalistic measures the AMA called for had been "imported from the dying despotisms of the Old Word": they were "un-American" and would destroy the republic whose role was to be "in the vanguard of progressive nations" (Spencer, 511). The NLMF had therefore history on its side. In Flower's Whig interpretation of American history, the League merely continued the long tradition of fighting for freedom—after the battle for political and religious freedom, the time had come to fight for medical freedom ("Menace of Medical Monopoly", 416). Privilege, in any form, was the enemy of "this sacred trinity" ("Menace of a National Health Bureau", 237).

Flower's teleological and exceptionalist vision of America was also theological. He framed his crusade in religious terms. He compared AMA doctors to the "clergy of the dark ages" who, out of jealousy because alternative schools of medicine were growing in popularity, raised the "convenient cry of prejudice and ignorance–quack, impostor, charlatan" in order to silence them ("Object Lesson", 507-10). The AMA was the modern equivalent of the medieval church infringing upon religious rights.

Following Herbert's Spencer's comments in *Social Statics*, Flower drew an analogy between "committing to government guardianship the physical health of the people, and committing to it their moral health." For Spencer and Flower, both the medical profession and the clergy in the Middle Ages were state-supported guardians of orthodoxy claiming they were defending the ignorant against heretics, while securing themselves government positions and power, in other words, "nine parts of self-interest gilt over with one part philanthropy" ("Spencer" 511). Flower's radically democratic, egalitarian, anti-elitist vision of authority took root in the Protestant tradition of an unmediated relationship between the individual and his spiritual, and here physical, welfare. No hierarchy–the church, the AMA or government–should come between people and their beliefs. The crusade for medical freedom was therefore the Reformation redux. As Flower put it: "it is God's battle we are fighting" ("Second General Conference", 10). It made sense for Flower to see medical freedom as both a scientific and religious crusade as, in his opinion, the Reformation, by enabling individuals to break free from medieval obscurantism and authoritarianism, gave birth to modern science. Americans had to live up to the American promise of freedom and to the historical promise of the Renaissance (*Century of Thomas More,* 289).

What Flower hated in medical laws was that they chained patients to one school of medicine and that they substituted compulsion for persuasion. An AMA physician even shamelessly admitted it in the *Journal of American Medical Association* in the June 8, 1907 issue: "compulsion, not persuasion is the keynote of State medicine" ("Compulsory Medical Inspection", 154). Flower believed authorities were not entitled to compel citizens to favor one type of medicine. By giving "special privileges" to the AMA, the government would leave patients with no choice. A modern approach would leave the different schools of medicine compete and win over the public with persuasive arguments. It would not use public authorities to push for a special agenda. Flower also objected to vaccination and school inspections since "compulsory inspection, to be effective, must be followed by compulsory treatment" by a school of medicine patients may object to (*First Report,* 17). Vaccination, for Flower, proved dangerous, as shown by numerous cases of gruesome deaths–"the slaughter of the innocents of the altar of serum therapy" as he put it ("Battle", May 1911 177).[5] It also violated two fundamental American principles–"the inviolability of the home" and "freedom from unwarranted search".[6] Another example of "invasion" was the controversial practice of stripping children for school inspections (Petrina, 217–218). Flower argued that, as Americans were intelligent individuals, citizens should be free to choose their physicians on the free market of ideas. Self-government could therefore be also understood in medical terms: in a *Medical Freedom* cartoon, Uncle Sam assures his American patient that he is historically and constitutionally "free to choose" among different treatments.[7] Flower refused the argument that state doctors should interfere with patients' choice. He dreaded compulsion, even for citizens' good: where would the government stop? Could citizens stop them once the precedent of vaccination was accepted? The argument of progressive Supreme Court Justice Oliver Wendell Holmes, Jr. in favor of compulsory sterilization in *Buck v. Bell* (1927) should give us food for thought and give some credit to Flower's concerns: "the principle that sustains compulsory vaccination is broad enough to cover cutting the Fallopian tubes. Three generations of imbeciles are enough."

The Politics of Expertise: Modernization vs. Antimonopoly

Not only did the controversy between the AMA and the NLMF oppose two narratives of reform and divided progressives but it also opposed two visions of science at a time when "science" was *the* legitimizing tool for intellectuals and politicians. Who was then to decide and impose what was real science or not? As Flower declared: "who is to decide who the medical charlatans are?" ("National Health", 512). Progressives in the 1910s disagreed: should it be the patients themselves—i.e. the "people"—or AMA experts working for a Department of Health? Flower was part of an old American tradition that saw science as a democratic activity and did not separate amateurs and experts into distinct spheres of competence. A growing class of scientists had increasingly contested that definition by the end of the 19th century and pushed for the modernization, standardization, and professionalization of their activity. They wanted to create a new role for themselves as "the guardians of the scientific light" by imposing a strict demarcation between genuine science and pseudo-science (Coon, 150). The AMA sought to elevate doctors into an autonomous and self-regulating elite community with strict boundaries. Alternative medicine was dismissed as unscientific, irrational and archaic and its doctors as "quacks." Physicians also shrugged off patients' claim that alternative medicine worked for them: according to them and their supporters in the mainstream press people were misled by the self-serving frauds of the patent medicine interests ("Imaginary Dangers", 12; "The AMA called a Trust", 1876).

With proper scientists in charge, the AMA argued that public health could be managed in a more efficient, modern and scientific way. The federal government and the AMA exemplify the technocratic strand within Progressivism*: modernizing and improving public health required a* shift toward a bureaucratic order, where boards of college-educated specialists would regulate health activities by deciding on scientifically-determined efficient and uniform standards. Fisher used the example of the Bureau of Labor to show the harmless benefits of expertise. This Bureau could investigate, and therefore improve, health conditions in factories. It had "no warrant to enter factories." This precedent showed that there was consequently no danger of "invasion of homes" by a Health Bureau. Another example was the Department of Agriculture. It never became a "farmers' trust" and, even though it gave modern expert advice, it never interfered with farmers' "freedom" to use the "moon theory" of cultivation if they wished for instance. Fisher also used the racialized scientific discourse of his time to disparage the beliefs of the NLMF. In a modern nation, believing in alternative medicine was nothing more than trusting "amulets, rings, charms and other 'fantastic remedies.'" This kind of primitivism ought only to be seen as a spectacle, like the Dresden International Hygiene Exhibition (Fisher *Memorial*, 74). Philanthropic foundations, a typical Progressive-era construction, funded that expertise. The Carnegie Foundation for example sponsored the Flexner report (1910), which promoted systematic licensing and standards in medical education. AMA doctor Abraham Flexner contrasted the non-scientific, *a priori* approach of "sectarian medicine" with a pragmatic definition of modern medicine: "to use knowledge procured in various ways in order to effect certain practical ends" (156). To ensure that the proper scientific methods prevailed, medical education had to be standardized. Physicians, like psychologists in their fight against spiritualists, wanted to "draw the line of demarcation between those who are of the profession and those who are not" (Petrina, 207; Coon, 143).[8]

The modernization narrative put forward by the AMA and the federal government, which is reminiscent of Wiebe's classic assessment of the Progressive Era as the triumph of a scientifically-managed bureaucratic order, clashed with Flower's antimonopoly brand of Progressivism and "populism of the body" (*Radical Middle Class,* 179).[9] Flower believed that the NLMF represented the "people" in their struggle against "medical monopolies" and the definition of science "political doctors" wanted to impose. His alternative definition of science and modernity relied in a deep-seated belief in the popular intellect, whereas the AMA, he charged, dismissed people's experience. Flower quoted William James who pointed out physicians' imperviousness to cure testimonies:

> But the regular profession stands firm in its belief that such experience is worthless and that the whole history is one of quackery and delusion. In spite of the rival schools appealing to experience, their conflict is much more like that of two philosophers or two theologies. Your experience, says one side to the other, simply isn't fit to count. ("Restrictive Medical Legislation", 789)

Flower trusted people's intelligence, which is why *the League relied on advertising campaigns because they* appealed directly to newspaper readers who could make up their mind on their own. Ads were a way to update the Common Sense tradition, which deeply influenced Flower. The Common Sense tradition cross-fertilized the American democratic ethos with Baconian science and the Scottish Philosophy of Common Sense. The Baconian inductive method relied on sense data and Common Sense, which had shaped university curricula in the 19th century ("Dugald Stewart"), implied that citizens, expert or not, could see the facts plainly, just as anyone with eyesight could see a tree. They could therefore reach an impartial conclusion by sharing their perspectives. True science consisted in the unbiased observation of the plainly observable facts of nature. As a result, this tradition did not oppose scientific and popular knowledge: scientific knowledge could be democratized, while being professionalized. Throughout the 19th century, the lyceum lecture circuit and local newspapers had popularized eminent scientists' discoveries, thus facilitating the extension of science to the laity (Hazen, 8-16; Noll, 216-38; Rosenfeld). It followed then that, in Flower's opinion, citizens could have a say when it came to medical science and, as a consequence, should be free to choose their physicians.

Scientific progress also implied freedom of inquiry. *Flower saw his crusade as an intellectual struggle against medical dogmas.* He spoke out against "compulsion in theoretical fields"–competition between various schools was a condition of scientific progress, whereas restrictive bills fostered medical censorship and hindered medical advances ("Compulsory Medical Inspection", 153). He invoked William James, who, in 1890, had opposed medical legislation in Massachusetts. James argued that scientific research was by nature tentative and therefore law should not freeze the endless process of investigation. As a consequence, legislatures should not take sides in medical controversies. For instance, mental healing–i.e. New Thought and Christian Science–should not be spurned but studied, according to James. His experience as a physician and as a patient also made him wary of the arrogance of "medical 'scientists.'" Their "pretensions" were not "false" but "premature": "the absurdity they may contain is not that of a frog trying to look like an ox, but rather that of the embryo alligator furiously snapping its harmless jaws, while yet enclosed by the egg-shell" (James, 227). Flower used the same arguments. Politicians should not interfere with the scientific process, just like doctors should not go into politics. He also contended that medicine was not an "exact science" but a "progressive art" (*Progressive Men,* 311-314; "Restrictive Medical legislation", 78§-90). History, for instance, had proved the dangers of heroic medicine like bloodletting or blistering. Contemporary experience showed the dangers of opium, which was widely used in medications, as well as the dangers of vaccination. He also kept quoting respected scientists who opposed the infringement of the state into medical matters: William James but also Benjamin Rush, Herbert Spencer, and Thomas Huxley.[10] It was necessary for Flower and James to maintain freedom of research. Critical thinking enriched by open-mindedness therefore should guide the scientific community. The AMA, on the contrary, was dogmatic and that meant returning to the "Dark Ages with its intolerance, prohibition, and class and creedal assumptions" (*Progressive Men,* 304).

Flower cast the medical controversies as an eschatological battle, a "civilization-moulding struggle" between "conservative, privilege-seeking restriction" and "progress under the aegis of freedom" (*Progressive Men,* 300). His typically progressive environmental determinism and his organicistic vision of a just, interdependent society as a healthy body framed his response to the AMA.[11] Different medical theories ought to circulate, like the air in model tenements or commodities in a free trade system. "Law-bulwarked privilege," whether state-sponsored medicine or tariff barriers, led to trusts and had "impoverishing and demoralizing effects" upon business, intellectual life and health (*Progressive Men,* 299). For instance, state-sanctioned professional arrogance did not stop regular doctors from incompetence. Flower's wife, for instance, had been dying from a disease no regular doctor could treat–they could not even agree on a diagnosis. A Christian Scientist practitioner cured her in two weeks ("Menace of Medical Monopoly", 411-14). Curing the nation therefore meant purifying society from those artificial barriers in order to restore the natural, free circulation of men and ideas. However, the major difference between the crusade for medical freedom and progressives' environmental reforms was that the body trope was no longer a metaphor. The machinations of the AMA endangered the concrete body of American citizens. Without rivals and divergent opinions, professional arrogance could turn into professional irresponsibility and kill. English astronomer Richard Proctor's death at the hand of New York physicians exemplified such "professional recklessness in a trust-ridden state" ("Menace of Medical Monopoly", 404).

Flower praised the scientific method but also questioned science's claims of infallibility. While mocking scientists' hubris, he was sensing a more modern approach to science William James and other pragmatists were developing, an approach Karl Popper would later call falsifiability, in other words, the possibility "for an empirical scientific system to be refuted" (18). Flower also shared with Pragmatists the idea that knowledge was tentative and collective as well as the anti-elitist belief that it "was accessible to the rank-and-file membership of an educated, democratic society" (Hollinger, 34-35). For James, Truth was "simply a collective name for verification-processes, just as health, wealth, strength, etc., are names for other processes connected with life, and also pursued because it pays to pursue them" (104). If mind-cure had indeed cured patients and therefore proved true, then mind-cure must correspond to something real in some people. Flower concurred–how else to explain the multiple cures effected by alternative medicine? That Flower could have affinities with pragmatism makes sense when we bear in mind that Charles Peirce also named it "critical common-sensism," a modern update on Scottish Common Sense (296-97, 305).

Flower feared the centralization of power that accompanied bureaucracy. He voiced what Hofstadter called "the complaint of the unorganized against the consequences of organization" (216), "the unorganized" being alternative doctors and their patients; "organization" the AMA. He dreaded that the bureaucratization of knowledge would challenge the authority of "the people," dispossess them of their decision-making power and place it in the hands of experts. Flower did not object to expertise *per se*. However, he was concerned over who would control experts. As he said: "we hold that sanitary activities should be under the direction of competent sanitary engineers, and not under the control of political doctors" ("B.O. Flower Answers", 11). Flower did not want AMA physicians to monopolize regularity powers, as it would mean they would "be parties to the very question to be decided by them" (*Federalist*). Such a conflict of interest was bound to drive out competitors from minority schools of medicine. Expert boards had to remain impartial. Health matters should therefore be left to "an unbiased Interstate Commerce Committee instead of a "doctors' committee" (Second General Conference", 8) and the way to ensure neutrality and fairness was to have all schools of medicine represented. The controversy over medical freedom opposed a managerial model based on regulation by professional organizations and government bureaucracies to an alternative model of democratic, deliberative professionalism, whereby "medical boards" would democratically represent all the medical schools and each school would confer degrees recognized by authorities because politicians could not and should not determine what is scientific or not. Flower also wanted specialists to bring out the facts but citizens to have the final say.[12] Lastly, scientific studies and policy implementation should be divorced. College-educated experts should not hold positions of power, nor should they administer the reforms—they should agitate for ideas in order to influence public opinion. The press, the "real congress and senate of the people," could then organize the democratic debate between experts, citizens and politicians and thus maintain the continuum between amateurism and expertise ("Notes", 2229). Flower refused the nascent policy-oriented, institutionalized partnership between federal government and higher education that was reorganizing knowledge and that would "engineer and manage a new America" (Zunz, xi).

Flower also feared that the head of a department of health could have the power to both execute and interpret the law, as had proved, according to Flower, "the persistent attempt in recent years of executive officials and departments and bureaus to arrogate legislative and judicial function" ("Chicago Conference", 3). Such arbitrary bureaucratic rulings would mean "handing the nation's health over to any bureau dominated by a monopoly-seeking special class," which would precipitate America into "Russian bureaucratic despotism" ("Menace of a National Health Bureau", 239). For him, it was no paranoia, as the Postmaster General had already arrogated to himself the right to exclude form the mails medical advertisements that he claimed were fraudulent. In 1902, the Supreme Court in School of Magnetic Healing v McAnnulty ruled that, the Postmaster General made "a fraud order depend on his opinion as to the efficacy of the complainant's method of *healing" (their emphasis).* The capacity of the mind to heal was "matter of opinion […] even intelligent people may, and indeed do, differ among themselves as the extent of this mental effect." The ruling concurred with Flower's skeptical attitude toward expert authority and his fear of arbitrary bureaucratic decisions. With a department head able to wield almost unlimited power and the AMA controlling the powerful machinery of government to further its vision of health and control school inspections, medical education, the certification of physicians, unproven and potentially dangerous vaccination campaigns and the issuance of death certificates, Flower thought the AMA was "seeking to gain complete control over the bodies of the people from the cradle to the grave" ("The Second Annual Meeting", 3).

Defining Science: Naturalistic Materialism vs. "Spiritual Idealism"

In Flower's Spencerian perspective, medicine had steadily evolved from lower and crude to higher and more civilized treatments. Heroic medicine had given way to more humane treatment like homeopathy, which, after an initial hostility, was finally co-opted by regular physicians. The upward trend toward safer and more civilized treatments was not over. The logical final stage of evolution was mental healing, and in particular Christian Science ("Menace of Medical Monopoly", 406; "Restrictive Medical Legislation", 801). The success of Christian Scientists and their persecution by regular physicians became a regular feature in *The Arena* in 1907 ("Recent Reckless and Irresponsible Attacks"; "The Seven Alleged Delusions"). Flower also published a book, *Christian Science as a Religious Belief and a Therapeutic Agent*, to defend them. It is their oppression that led to the resurgence of the crusade for "medical freedom" that he had begun in the 1890s.

Flower believed the premise of AMA science to be wrong and archaic. He opposed, on the one hand, the materialist worldview of modern, naturalistic science that was gradually being established and imposed top-down by professional societies, like the AMA,[13] and, on the other hand, what he called "spiritual idealism," in other words, science that would not shy away from investigating the uncharted territories of the invisible, in particular the powers of mind. Flower believed that mind could act on the body, hence his interest in mental therapeutics. "Spiritualizing" health issues, so to speak, Flower could not but be appalled at vaccination and school inspections. The power of the mind was potentially infinite and he accused inspectors of "the evil of creating fear in the child mind," which could literally make them sick ("Compulsory Medical Inspection", 153).[14] Flower's reform agenda included remoulding people's environment to solve the problems of modern, industrial cities (with municipal housing, parks, playgrounds, ventilation etc.) but also remoulding public opinion by exposing AMA and government propaganda while publicizing "thought-molders." He took the term literally. For him, brains were "plastic"—they were susceptible to influence and could be modified and "cured." A healthy citizenry implied the circulation of healthy ideas "that may infect others" as he put it: "thought is contagious and people are thinking" (*Progressive Men*, 22, 153; "Transition", 124). The uplifting power of his magazine work could therefore stop the disease-breeding scaremongering of the AMA.

Since Flower believed mental healing and alternative medicine in general were a more evolved form of medicine, he contended that materialistic doctors were incompetent to judge their value, in the same way that believers in the Ptolemaic theory could not judge the more advanced Copernican theory. For Flower, alternative medicine reflected a sort of "paradigm shift"–a revolution in science–and the two paradigms were incommensurable (Kuhn 6, 103). No matter what sort of *artificial* bureaucratic organization the AMA could devise to impose its worldview, mental therapeutics would eventually prevail since *natural* evolution and history were on its side; in other words, even though the AMA and the federal government managed to impose top-down regulations of medical practice, their authority remained *ipso facto* meaningless. The AMA, with its modernization narrative, used the same kind of biologized and Whig approach to history: the US had to adapt to its modern environment. Since both sides could see their opinion as *natural* and *scientific*, the debate could not be settled from an intellectual standpoint, hence its displacement in the political arena. Expertise became rhetoric and undermined science's pretense to objectivity. Debates did not revolve around the truth of arguments anymore: they were epistemological (which knowledge is valid in a modern society) and legalistic (advancing facts and evidence to expose the invalidity of opponents' arguments).

Is Public Health a Political, Scientific or Religious Issue?

Flower saw health problems as first and foremost an economic and class problem. Health issues could be tackled more effectively with more justice, not more bureaucratic centralization. For him, AMA reforms "divert[ed] the public mind from great economic wrongs." They were "palliative treatment advanced by a privileged class in lieu of fundamental remedies." He wanted government to treat the causes, not the symptoms, of health problems. These remedies consisted in economic reforms to fight against child labor, poverty and slums, in other words, "wise general legislation that shall make for economic independence and health through wholesome environment," not an "army of tax-supported state doctors." Visionary architects, for instance, were more needed that physicians or school inspections ("Compulsory Medical Inspection", 153-60). A case in point was what happened in New Zealand and Australia. Flower attributed the success of their public health policy to the advanced and humanistic antitrust legislation, not their Departments of Health. They broke up monopolies–by ensuring free and equal access to the common resources–and treated the causes of inequalities and, therefore, of health problems ("Menace of a National Health Bureau", 240-41).

Flower also wanted to replace compulsory, top-down, government-backed, and potentially dangerous, health campaigns like school inspections and vaccination with sanitation at the local level, where democratic debate could control health procedures, and with "general instruction in regard to the importance of pure air, water, sunshine, exercise and food–questions about which there are no warring opinions" ("Compulsory Medical Inspection", 158). He praised, for instance, what Cleveland progressive mayor Tom L. Johnson had done in his city to stamp out smallpox. He put an end to vaccination since it was "sowing the seeds of disease" and, instead, conducted a harmless and very effective house-to-house campaign of disinfection ("How Cleveland"). Work was more effective at the state and municipal level, whereas centralization would remove all sense of responsibility and weaken efficiency. AMA reforms like the Owen bill, with more government officials, would also mean an increase in taxation ("Rising Revolt", 5).

Flower's position was ambiguous, as he both feared the encroachment of the state into private beliefs while calling for the expansion of federal power to achieve social and economic justice. For Fisher, Flower "believe[d] in the outworn doctrine of laissez-faire," which would have surprised Flower as he championed public ownership of railroads for example. He did not mind government intervention to regulate the social, political and economic ills caused by excesses of industrialization but he refused any interference with personal beliefs. Fundamental reforms, for him, only tampered with people's environment, not their opinions or their bodies. Flower summed up his political philosophy like this:

> We are of those who would favor the state going far–very far–to remedy the wrongs that work injury to the weak and unfortunate, especially when the evils affect the young; but we unhesitatingly oppose the fastening on the body politic of representatives of a favored school of medicine. ("Compulsory Medical Inspection", 158)

When it came to personal beliefs, and he saw health as a sort of religious issue, Flower remained rooted in an older "libertarian radicalism" that "interposed personal sovereignty and rationality in social, religious, sexual, [and medical] spheres against the power of church and state" (Rabban, 53). In a way, Flower's crusade could be seen as the struggle of radical democrats against what Michel Foucault called "biopower," in other words, the "subjugations of bodies and the control of populations" thanks to state-controlled medicine and the power to determine which knowledge is valid in a modern society (140).

A cabinet-level Department of Health was only created in 1953. On that point, Flower was successful. Of course, he did not stop the professionalization of modern medicine. Other ironies abound in the story of the NLMF. Its president, Flower, had stopped all involvement in its activities in 1915, before the League disbanded in September 1916: in a language reminiscent of his crusade against medical censorship, he now battled another hierarchical organization trying to censor citizens–the Catholic Church which used the Comstock laws to censor supposedly "obscene" anti-Catholic books and newspapers ("Story", 3-10). Nowadays, alternative medicine has been co-opted, and arguably taken over, by "regular physicians," as shown by the establishment of the Office of Alternative Medicine in 1991 and of the National Center for Complementary and Alternative Medicine in 1998. Flower's ambivalent radical position–culturally libertarian but economically, politically and socially on the left–forces us to rethink traditional political categories. Recent medical controversies have again shown that the reformist/conservative dichotomy proves inadequate when it comes to "the politics of healing." In 1992, both Senators Tom Harkin, an "extremely liberal agricultural rebel," and Orrin Hatch, "an extremely conservative Mormon free-enterpriser," sponsored the US Office of Alternative Medicine. Catholic activists, like Phyllis Schlafly, or a liberal Jewish intellectual, like Michael Lerner, also promoted alternative medicine. Flower conflated these "ideological non-bedfellows" in one figure (*The Politics of Healing,* 1-6). He meant "medical freedom" to be a bulwark of Progressivism. However, and ironically enough, conservative Republicans use the same arguments today to defeat liberal health legislation. For example, the opponents of Obamacare often invoked the principle of "medical freedom." In June 2014, the Supreme Court, in Burwell v. Hobby Lobby, ruled that corporations owned by religious families could not be required to pay for contraception coverage for their female employees. This decision, in the words of the *New York Times*, "opened the door to many challenges from corporations over laws that they claim violate their religious liberty." This would have deeply disturbed Flower because for him religious and medical freedoms were supposed to help reformers defeat trusts.

The demise of the Common Sense tradition and the rising dominance of a more bureaucratized and technocratic vision of science at the beginning of the 20th century had unintended political and intellectual consequences. Flower's libertarian and antimonopoly radicalism dovetailed with the Common Sense notion that citizens were intelligent enough to understand the world. Patients could, and therefore should, govern themselves medical-speaking: it was their rights as Americans, their individual reasonability as members of the community and the key to individual and collective progress. Government should therefore facilitate the free and equal access of all schools of medicine to open public debate and ensure them a level-playing field. With the triumph of "modern" science and the technocratic approach to health, Common Sense philosophy and antimonopoly politics were banished beyond the pale of "Progressivism" and later seized on by conservatives. As a result, nowadays Flower's rhetoric is used to different ends. Medical freedom advocates today use the same premise (the intellectual capacity of patients to decide which treatment is best for them and therefore the moral and political imperative to let them choose) but public authorities should stay out of the picture. Big government *in itself* is seen as a threat to democracy, whereas Flower only condemned its corruption and takeover by special interests. For Flower, government should empower individuals and guarantee that they could exert their responsibility, while for the opponents of "socialized medicine" individual responsibility excludes government intervention altogether. As to technocratic progressives (or liberals as we would say today), they think that government in partnership with medical science has a role to play to promote public health and that only professionals can make an enlightened decision about the proper treatment for the patient. Individual responsibility is reduced to choosing a healthy lifestyle, not much of a choice for Flower who believed that patients should also be agents of their own welfare.

Bibliography

Primary sources

"100,000 petition signers in ten days by advertising." *Printers' Ink* 71 (2 June 1910): 47-49.

Adams, Samuel Hopkins. 1906. *The Great American Fraud, Articles on the Nostrum Evil and Quacks, in Two Series, reprinted from Collier's Weekly*. New York: P. F. Collier & Son.

"The American Medical Association called A Trust". 1910. *Journal of the American Medical Association* 54.23 (June 4). [1876]

American School of Magnetic Healing v. McAnnulty. 187 U.S. 94. Supreme Court of the US. 1902. Justia (n.d). To be consulted at: https://supreme.justia.com/cases/federal/us/187/94/.

Accessed 3 Jan. 2015)
"A Bad Bunch." *Collier's* 6 May 1911, p. 10.
"Beware, O Ye Doctors!" *New-York Tribune* 16 May 1910, p. 12.
Briggs, Finney. 1914. "A the people want it." *Medical Freedom* 4 (Dec. 1914), p. 57.
Buck v. Bell. 274 U.S. 200. Supreme Court of the US. 1927. *Supreme Court Collection.* Legal Information Inst. Cornell U Law School (n.d.) To be consulted at: http://www.law.cornell.edu/supremecourt/text/274/200. Accessed 2 Jan. 2015.
Chronicling America. Historic American Newspapers. Library of Congress (n.d). To be consulted at: http://chroniclingamerica.loc.gov. Accessed 1 Aug. 2014.
"The Closing of the League." *Medical Freedom* 6 (Sept. 1916), p. 2.
Congressional Record 47 (6 July 1911), pp. 2659-75.
"The Daily Lesson." *Medical Freedom* 5 (Oct. 1915), p. 27.
"Fight 'Doctors' Trust.'" *The New York Times* 10 January 1912.
Fisher, Irving. 1909. *National Vitality: its Wastes and Conservation.* Washington D.C.: Government Printing Office.
—. 1910a. "Defines Opposition to the Owen Bill." *The New York Times* 18 June 1910: 8; reprinted as "The Layman's Reply to the Medical Freedom (?) League." *Journal of the American Medical Association* 54.5 (June 18), pp. 2133-34.
—. 1910b. "A Department of Dollars vs. a Department of Health." *McClure's Magazine* 35 (July), pp. 329-30.
—. 1912. *Memorial Relating to the Conservation of Human life as Contemplated by Bill (S. 1) Providing for a United States Public Health Service.* Washington D.C.: Government Printing Office.
—. 1913. "National Health and Medical Freedom. Both Sides of a Question of Public Interest." *Century Magazine* 85 (February), pp. 513-14.
Flexner, Abraham. 1910. *Medical Education in the United States and Canada: A Report to the Carnegie Foundation for the Advancement of Teaching*. New York: Carnegie Foundation for the Advancement of Teaching.
Flower, Benjamin O. 1890a. "Notes on Living Problems of the Hour. Liberty of Citizenship Imperiled. A Consensus of Opinion on a Recent Notable Case. Introductory Statement by the Editor" *The Arena* 2 (July), pp. 229-30.
—. 1890b. "Praise for the Arena." *The Arena* 2 (November): n. pag.
—. 1890c. "A Transition Period." *The Arena* 3 (December), pp. 124-25.

—. 1891a. "An Object Lesson in Freedom." *The Arena* 3 (March), pp. 507-10.

—. 1891b. "Class Interests and the Rights of the People." *The Arena* 3 (March), pp. 510-11.

—. 1891c. "Herbert Spencer's Arraignment of Class Medical Laws." *The Arena* 3 (March), pp. 511-12.

—. 1898. "The Menace of Medical Monopoly" *The Arena* 9 (February): 400-16. Reprinted twice: "Restrictive Medical legislation and the Public Weal." *The Arena* 19 (June 1898), pp. 781-809, and "Medical Monopoly the Foe to Progress and Human Rights." *The Twentieth Century Magazine* 3 (Nov. 1910), pp. 148-56.

—. *The Century of Sir Thomas More*. 1896. Boston: The Arena Publishing Company.

—. 1902. "How Cleveland Stamped Out Smallpox" *The Arena* 27 (April), pp. 426-29.

—. 1907a. "The Recent Reckless and Irresponsible Attacks on Christian Science and its Founder, with a Survey of the Christian-Science Movement." *The Arena* 37 (January), pp. 47-67.

—. 1907b. "The Seven Alleged Delusions of the Founder of Christian Science Examined in the Light of History and Present-Day Research." *The Arena* 38 (October), pp. 408-22.

—. 1910a. *Christian Science as a Religious Belief and a Therapeutic Agent*. 2nd ed. Boston: Twentieth Century Company.

—. 1910b. "The Menace of a National Health Bureau." *The Twentieth Century Magazine* 2 (June), pp. 234-42.

—. 1910c. "Why We Oppose the Bills to establish a National Department or Bureau of Health." *The Twentieth Century Magazine* 2 (July), pp. 340-48.

—. 1910d. "Medical News Notes. The Fight for Medical Freedom." *The Twentieth Century Magazine* 2 (September), pp. 366-68.

—. 1910e. "Medical Monopoly the Foe to Progress and Human Rights." *The Twentieth Century Magazine* 3 (November), pp. 148-56.

—. 1910f. "Medical News Notes. The Revolt Against the Medical Hierarchy." *The Twentieth Century Magazine* 3 (December), pp. 271-73.

—. 1911a. "The Compulsory Medical Inspection of School Children." *The Twentieth Century Magazine* 4 (May), pp. 151-60.

—. 1911b. "The Battle for Medical Freedom." *The Twentieth Century Magazine* 4 (May), pp. 177-79.

—. 1911c. "The Chicago Conference." *Medical Freedom* 1 (December), pp. 2-3.

—. 1912a. "Try to Stop Free Speech!" *Medical Freedom* 1 (February), p. 2.
—. 1912b. "The Second Annual Meeting." *Medical Freedom* 1 (June), p. 3.
—. 1912c. "B.O. Flower Answers Irving Fisher." *Medical Freedom* 2 (Sepember), p. 11.
—. 1912d. "An Appeal." *Medical Freedom* 2 (October), pp. 2-3.
—. 1913a. "National Health and Medical Freedom. Both Sides of a Question of Public Interest." *Century Magazine* 85 (February), pp. 512-13.
—. 1913b. "The Second General Conference of the League. Address of Welcome by B.O. Flower." *Medical Freedom* 2 (August), pp. 8-10.
—. 1914a. *Progressive Men, Women, and Movements of the Past Twenty-five Years*. Boston: The New Arena.
—. 1914b. "Position of Political Doctors is Fundamentally Unsound." *Medical Freedom* 3 (July), p. 7.
—. 1914c. "An Earnest Word to Friends of Medical Freedom in Grave Crisis." *Medical Freedom*, 4 (October), pp. 25-26.
—. 1914d. "Rising Revolt Against the Burdens of State Medicine." *Medical Freedom* 4 (December), p.5.
—. 1915. "A Bird's-Eye View of the National league for Medical Freedom." *Medical Freedom* 4 (April), p. 117.
—. 1916. *Story of the Menace Trial: A brief sketch of this historic case with reports of the masterly addresses by Hon. J.L. McNatt and Hon. J.I. Sheppard, attorneys for the defense*. Aurora, MO: United States Publishing Co.
Flower Richard C. 1888. *The Allopathic Czar*. Boston: Massachusetts Medical Liberty League.
Hamilton, Alexander or Madison, James. 1788. "The Federalist Papers: No. 49. Method of Guarding Against the Encroachments of Any One Department of Government by Appealing to the People Through a Convention." *The New York Packet*. Feb. 5. The Federalist Papers. Library of Congress (n.d.). To be consulted at: http://thomas.loc.gov/home/histdox/fed_49.html. Accessed 2 Jan. 2015.
"Imaginary Dangers Proclaimed." *The New York Times* 27 Sept. 1910: 12.
Jacobson v. Commonwealth of Massachusetts. 197 U.S. 11. Supreme Court of the US. 1905. Justia (n.d). To be consulted at: https://supreme.justia.com/cases/federal/us/197/11/. Accessed 3 Jan 2015.
James, William. 1987. "Review of 'Rapport sur les progrès et la marche de la physiologie générale en France' by Claude Bernard (1868)." *Essays, Comments, and Reviews*. Cambridge, MA: Harvard UP.

Knopf, Sigard Adolphus. 1910. "The Owen Bill for the Establishment of a Federal Department of Health, and its Opponents." *Popular Science Monthly* 77 (October), pp. 373-78.

"Liberty." *Collier's* 3 June 1911, p. 9.

Liptak, Adam. 2014. "Supreme Court Rejects Contraceptives Mandate for Some Corporations. Justices Rule in Favor of Hobby Lobby" *The New York Times* 30 June. To be consulted at: http://www.nytimes.com/2014/07/01/us/hobby-lobby-case-supreme-court-contraception.html. Accessed 3 Jan 2015.

Medical Freedom 1 (Sept. 1911), p. 2.

—. 2 (Oct. 1912), p. 2.

"Medical Freedom." Editorial. *The New York Times* 18 May 1910, p. 10.

The National League for Medical Freedom. 1910. *First Report of The National League for Medical Freedom Issued to its Members,* August. New York: National League for Medical Freedom.

"A New Combination Against the American Medical Association." *Journal of the American Medical Association* 54.22 (May 28, 1910), p. 1791.

Newspaper.com. Historical Newspapers from 1700s-2000s. (n.d). To be consulted at: http://www.newspapers.com. (last accessed 1 Aug. 2014)

"The Organization of Ill Health." *Journal of the American Medical Association* 55.5 (July 30, 1910): 421.

Peirce, Charles Sanders. 1935. "Issues of Pragmaticism." *The Monist* 15.4 (Oct. 1905): 481-99. *Collected Papers of Charles Sanders Peirce. Volumes V and VI: Pragmatism and Pragmaticism and Scientific Metaphysics.* Eds. Charles Hartshorne, Paul Weiss. Cambridge, MA: Harvard UP, pp. 438–63.

Roosevelt, Theodore. 1908. "Eighth Annual Message." 8 Dec. Gerhard Peters and John T. Woolley. *The American Presidency Project* (n.d.) To be consulted at: http://www.presidency.ucsb.edu/ws/?pid=29549. Accessed 2 Jan. 2015)

Taft William H. 1909. "First Annual Message." 7 Dec. Gerhard Peters and John T. Woolley. *The American Presidency Project* (n.d.) To be consulted at: http://www.presidency.ucsb.edu/ws/?pid=29550. Accessed 2 Jan. 2015.

—. 1910. "Second Annual Message." 6 Dec. Gerhard Peters and John T. Woolley. *The American Presidency Project* (n.d.) To be consulted at: http://www.presidency.ucsb.edu/ws/?pid=29551. Accessed 2 Jan. 2015.

"Who is Right? *Medical Freedom* 1 (March 12), p. 10.

Secondary sources

Coon, Deborah J. 1992. "Testing the Limits of Sense and Science: American Experimental Psychologists Combat Spiritualism, 1880–1920." *American Psychologist* 47.2: 143-51.

Cunliffe, John. 2000. "Introduction: Left-Libertarianism—Historical origins." *The Origins of Left-libertarianism: An Anthology of Historical Writings*. Eds. Peter Vallentyne and Hillel Steiner. New York: Palgrave, pp. 1-20.

"Dugald Stewart 1753-1828." International Association for Scottish Philosophy (n.d.) To be consulted at: http://www.scottishphilosophy.org/dugald-stewart.html. Accessed 31 March 2014.

Foner, Eric. 1994. "The Meaning of Freedom in the Age of Emancipation." *Journal of American History* 81. 2 (September), pp. 435-60.

—. 1999. *The Story of American Freedom*. New York: W.W. Norton.

Foucault, Michel. 1990. *The History of Sexuality, Vol. 1: An Introduction*. (translated from the French). New York: Vintage books. [1976]

Friedman, Milton and Rose. 1980. *Free to Choose. A Personal Statement*. New York: Avon Books.

Hazen, Craig James. 2000. *The Village Enlightenment in America: Popular Religion and Science in the Nineteenth Century*. Urbana: University of Illinois Press.

Historical Highlights. U.S. Department of Health & Human Services, n.d. To be consulted at: http://www.hhs.gov/about/hhshist.html. Accessed 3 Jan. 2015.

Hofstadter, Richard. 1955. *The Age of Reform : from Bryan to F.D.R.* New York: Vintage Books.

Hollinger, David A. 1989. "The Problem of Pragmatism in American History." *In the American Province: Studies in the History and Historiography of Ideas*. Baltimore: Johns Hopkins UP, pp. 23-43.

Johnston, Robert D. (ed.). 2004. *The Politics of Healing: Histories of Alternative Medicine in Twentieth-Century North America*. New York: Routledge.

—. 2003. *The Radical Middle Class: Populist Democracy and the Question of Capitalism in Progressive Era Portland, Oregon*. Princeton: Princeton UP.

Kuhn, Thomas S. 2012. *The Structure of Scientific Revolutions*. 4th ed. Chicago: U of Chicago Press. [1962]

Marin-Lamellet, Jean-Louis. 2014. "What is the cocoon but a dark cabinet?': Benjamin O. Flower, Print Culture and the Legitimisation of

Fringe Science in the 1890s." *Mémoires du livre / Studies in Book Culture* 6.1 (fall 2014): n. pag. To be consulted at : http://www.erudit.org/revue/memoires/2014/v6/n1/1027697ar.html. Accessed 3 Jan. 2015.

Noll, Mark A. 1985. "Common Sense Traditions and American Evangelical Thought" *American Quarterly* 37.2 (1 July), pp. 216-38.

Petrina, Stephen. 2008. "Medical Liberty: Drugless Healers Confront Allopathic Doctors, 1910–1931." *Journal of Medical Humanities* 29.4 (Aug. 7), pp. 205–230.

Popper, Karl. 2002. *The Logic of Scientific Discovery*. London, New York: Routledge Classics. [1935]

Rabban, David M. 1992. "The Free Speech League, the ACLU, and Changing Conceptions of Free Speech in American History." *Stanford Law Review* 45.1 (Nov. 1), pp. 47-114.

Rosenfeld, Sophia A. 2011. *Common Sense: A Political History*. Cambridge, MA: Harvard UP.

Vallentyne, Peter. 2000. "Left-Libertarianism: A Primer." *Left Libertarianism and Its Critics: The Contemporary Debate*. Eds. Peter Vallentyne and Hillel Steiner. New York: Palgrave, pp. 1-20.

Wiebe, Robert. 1967. *The Search for Order, 1877-1920*. New York: Hill and Wang.

Willrich, Michael. 2011. *Pox: An American History*. New York: The Penguin Press.

Zunz, Olivier. 2000. *Why the American Century?* Chicago: University of Chicago Press.

Notes

[1] Flower edited *The Arena* in Boston between 1889 and 1909 (with an interruption between 1896 and 1900), *The New Time* (Chicago) in 1897 and 1898, *The Coming Age* (Boston) in 1899 and 1900, *The Twentieth-Century Magazine* (Boston) between 1909 and 1911, *Medical Freedom* (Chicago) between 1912 and 1915 and *The Menace* (Aurora, MO) between 1916 and 1917.

[2] The official journal of the League, *Medical Freedom*, and the *Journal of the American Medical Association* provide a comprehensive view on both sides of the controversy.

[3] In the 1910s, the Marine-Hospital Service was the only federal organization dealing with public health issues. At the end of the 19th century, it had increasingly taken over quarantine functions from the States. In 1902, the Marine-Hospital Service had been renamed the Public Health and Marine Hospital Service "in recognition of its expanding activities in the field of public health," notably medical inspection of arriving immigrants (*Historical Highlights*). A federal

Bureau of Health would unite the Marine-Hospital Service, "the police jurisdiction of the Federal Government respecting quarantine," as well as the research activities conducted by "competent experts into questions of health affecting the whole country, or important sections thereof, questions which, in the absence of federal governmental work, are not likely to be promptly solved" (Taft, "Second Annual Message").

[4] NLMF advert. "Do You Want the 'Doctors' Trust' to be able to Force its Opinions on You?" *Washington Times* 19 May 1910: 6. *Chronicling America: Historic American Newspapers*. Library of Congress. http://chronicling america.loc.gov/lccn/sn84026749/1910-05-19/ed-1/seq-6/. 2 Jan. 2015.

[5] Historians and doctors have often dismissed anti-vaccinationists as ignorant and dangerous "charlatans." However, they judge according to their own criteria, which is the crux of the problem for Flower as he challenges the AMA's scientific approach. Scientifically speaking, things at the beginning of the 20th century were not as straightforward as today: the victory of the germ theory remained uncertain and the vaccine used then, produced from infected cows and carried out without any safety regulations, was an unreliable and far from a risk-free treatment: for example, even vaccinated people could have smallpox. On top of that, because of the absence of regulation, cheap vaccines could leave patients disabled (safety issued remained a problem until the mid 20th century despite governmental control; even today, some scientists admit the risks). There was also a strong class dimension in vaccination at the time: government doctors targeted poor immigrants and vaccination "raids" were extremely violent. Natives in colonies resented it as yet another tool of imperialist control, just like labor activists saw it as a form of social control. Given the risks, the absence of scientific consensus and Flower's radical sympathies, the refusal of vaccination could therefore be seen as a rational choice at the time (*Radical Middle Class,* 179-80; Willrich, 1-6, 214-15).

[6] For example, the cover of the May 1916 issue of *Medical Freedom* shows the fall of the house of America. The country is visualized as a neoclassical temple and each pillar represents a fundamental right. The first two columns ("the inviolability of the home" and "freedom from unwarranted search") are destroyed, thus jeopardizing the whole structure. "What Medical Legislation Would Do." *Medical Freedom* 5 (May 1916). *HathiTrust's digital library*. http://babel.hathitrust.org /cgi/pt?id=mdp. 39015013773752;view=1up;seq=327. 2 Jan. 2015

[7] Briggs, Finney. "As the people want it." *Medical Freedom* 4 (Dec. 1914): 57. *HathiTrust's digital library*. http://babel.hathitrust.org/cgi/pt?id=mdp.39015013 773752;view=1up;seq=63. 2 Jan. 2015 . "Free to choose" will remind the 21st century historian of economists Milton and Rose Friedman's 1980 book and television series that advocated free market principles and laissez-faire economic policies. However, we should be careful not to read back late 20th-century debates over government interventionism and the rising influence of libertarianism in right-wing circles. As Foner's *Story of American Freedom* demonstrates, the very meaning of "freedom" has been contested throughout American history and should not be confined to the definition that Friedman and other conservative intellectuals have, arguably, imposed in American debates in the last quarter of the 20th century.

In the 19th century, libertarians did push for progressive social reforms such as radical abolitionism, free love (i.e. the right not to be enslaved to your spouse if you love somebody else, the right to divorce in short), and freethought (Rabban, 53). Economically speaking, left-libertarianism, even though it may be nowhere to be found nowadays, did attract major thinkers in the past, like Tom Paine or Henry George. According to them, individuals own themselves and thus "owe no service the others expect as the result of voluntary action." Natural resources, on the other hand, are owned by the community, and may be appropriated only with their permission, or with compensation; in other words, "self-ownership and private property in artifacts must be combined with an initially egalitarian entitlement to raw natural resources" (Cunliffe, 1-5; back cover). Taxations and rents, like Henry George's Single Tax, should compensate for the appropriation of natural resources; the "social fund" generated should be spent so as to "promote equality of opportunity for well-being" and "to improve the opportunities for a good life of those with relatively unfavorable genetic or situational endowments, and spends little or nothing (except for instrumental reasons) on those with relatively favorable endowments" (Vallentyne, 10-11).

[8] In the 1890s, Flower also sought to legitimize this other type of "fringe" science: spiritualism, or as he called it "psychical research" (Marin-Lamellet).

[9] According to Wiebe, progressives were middle-class reformers who sought to bring order to the social chaos caused by rapid industrialization. They replaced the values of small-town community life with a "bureaucratic orientation" based on "the values of continuity and regularity, functionality and rationality, administration and management" (Wiebe, 295).

[10] Flower often supplied tables in two columns, contrasting quotes by illustrious figures who championed medical freedom on the left (William James, Herbert Spencer, Thomas Huxley, John Stuart Mill, Benjamin Franklin etc.) and obscure AMA physicians who supported state medicine on the right. "Who is Right?" *Medical Freedom* 1 (March 12): 10. *HathiTrust's digital library.* http://babel.hathitrust.org/cgi/pt?id=mdp.39015038695758;view=1up;seq=86. 2 Jan. 2015.

[11] Progressives thought that individuals were the product of their environment. The least fortunate were not to blame for social ills like poverty, criminality and vice as they could not develop properly in unhealthy surroundings and were contaminated by their poisonous atmosphere. As a result, reforming meant first and foremost to modify the environment in order to create a breeding ground for a healthy citizenry.

[12] That is why Flower also favored direct democracy and cooperatives, in other words, organizations controlled by the people.

[13] According to the OED, this is "the most usual sense since the mid-19th cent. when used without any qualification."

[14] In a cartoon, a teacher-like physician is at school. He scares a pupil with a fishing rod with a skull at the end of the line. He also points to a blackboard that reads, "Name the ten most frightful diseases and describe them. Explain why the very air you breathe is full of deadly germs. Tell why children should be inoculated

for everything." "The Daily Lesson." *Medical Freedom* 5 (Oct. 1915): 27. *HathiTrust's digital library*. http://babel.hathitrust.org/cgi/pt?id=mdp.39015013773752;view=1up;seq=225. 2 Jan. 2015.

SCIENCE AND ANIMAL ETHICS: PETER SINGER'S UTILITARIAN APPROACH

ÉMILIE DARDENNE

The second half of the past century has seen many changes in the perception and use of domesticated animals. Their exploitation in intensive farming and research has skyrocketed,[1] while animal rights theory has developed and become prominent in ethical debates. In a discourse increasingly imbued with ethological and biological notions,[2] stemming from progress made by scientists working on animal behaviour and capacities, animal advocates have suggested that animals should be entitled to more rights, and that there might not be anything so distinctive about humanity that humans should have moral status and nonhumans should not. This ethical issue has kept philosophers busy for centuries. Indeed, while many have striven to establish the uniqueness and superiority of human beings,[3] others have adopted a different epistemological approach, summoning scientific ideas to shed light on the possible limits of anthropocentrism, and arguing for a reappraisal of the status of animals.[4]

In this paper, I will look at the work of one of these philosophers, Peter Singer. Through his utilitarian approach developed in *Animal Liberation* (1975), *Practical Ethics* (1980) and in collective works such as *The Great Ape Project* (1993), he has argued, thanks to developments in biology, zoology, ethology, psychology, and neurobiology, that animals are individuals, that they are sentient beings who have needs and interests which should be equally considered. As Singer claims,

> there are no good reasons, scientific or philosophical, for denying that animals feel pain. If we do not doubt that other humans feel pain we should not doubt that other animals do so too. Animals can feel pain [...] there can be no moral justification for regarding the pain (or pleasure) that animals feel as less important than the same amount of pain (or pleasure) felt by humans. (1995, 15)

Singer has brought detailed philosophical and scientific information together to enlarge a vision of ethics, and to pursue a political and a legislative agenda based on a less humanistic and more utilitarian perspective. Paying close attention to the manner in which the philosopher articulates scientific views to the ethical models he advocates, this paper will examine how he moves from scientific facts to a normative approach, and will also consider the context within which Singer argued for equal consideration of interests. I will first show how Singer has summoned scientific facts to advocate his principle of equal consideration of interests based on sentience. I will then tackle the different types of criticism levelled at Singer: one claims that he goes too far in reappropriating scientific concepts and data, and that his views are "an affront to our common humanity" (Elisabeth de Fontenay); another claims that he does not go far enough and should argue for a more robust protection of the environment as a whole, not just animals (Holmes Rolston III).

Singer's Argument: from Species to Sentience

Peter Singer suggests that we follow Jeremy Bentham and settle upon sentience, the capacity to experience pleasure and pain, as the relevant qualification for moral consideration. In his work on animal ethics he defends the view that all sentient beings are worthy of moral attention, and that no special status comes from being a member of the human species. In so doing, he moves the biological cursor defining the contours of our ethical world from the boundary of the human species to the notion of sentience:

> The capacity for suffering and enjoying things is a prerequisite for having interests at all, a condition that must be satisfied before we can speak of interests in any meaningful way. […] If a being suffers, there can be no moral justification for refusing to take that suffering into consideration. No matter what the nature of the being, the principle of equality requires that the suffering be counted equally with the like suffering—in so far as rough comparisons can be made—of any other being. If a being is not capable of suffering, or of experiencing enjoyment or happiness, there is nothing to be taken into account. This is why the limit of sentience (using the term as a convenient, if not strictly accurate, shorthand for the capacity to suffer or experience enjoyment or happiness) is the only defensible boundary of concern for the interests of others. (2011, 57)

The yardstick for Singer's "equal consideration of interests" is sentience. Three arguments support the idea that nonhuman animals can feel pain as much as humans do:

- The behaviour of a being that suffers: whether he or she is human or animal, the signs are identical.
- The physiological argument of the nervous system: both humans and animals possess one.
- The evolutionary usefulness of pain.[5]

In terms of sentience, humans and animals are equal[6] and what makes a being important is the fact that he/she can feel pleasure or pain. This is when Singer seizes upon scientific ideas:

> It is surely unreasonable to suppose that nervous systems that are virtually identical physiologically, have a common origin and a common evolutionary function, and result in similar forms of behavior in similar circumstances should actually operate in an entirely different manner on the level of subjective feelings. […] The overwhelming majority of scientists who have addressed themselves to this question agree. Lord Brain, one of the most eminent neurologists of our time, has said: "I personally can see no reason for conceding mind to my fellow men and denying it to animals […]."[7] (11-12)

Defining the ability to suffer as a critical property, and claiming that all sentient beings are worthy of moral attention implies nothing less than an ethical revolution, according to Singer. The status of animals is elevated, so that they now have moral standing, on a par with all humans, including those who share what have traditionally been considered as animal deficiencies (deficiency in language, autonomy, self-awareness, intentionality, and a sense of existing over time, etc.). Because they are sentient, animals have interests. Because they have interests, they should count morally.

Now, if sentience is enough to make animals count in their own right, it is not enough, as a criterion, to consider that they should not be killed. Killing a sentient being and causing him/her suffering do not bear the same moral significance.

The assessment of a pain depends on how intense it is and how long it lasts, Singer writes, but pains of the same intensity and duration are equally bad, whether felt by humans or animals. When the value of life is considered, Singer goes on, we work at a whole other level. We have to consider whether the being is self-aware and sees himself/herself as existing over time, or if he/she is merely conscious. If a being is not self-aware, death cannot be contrary to an interest in continued life, so this being can be killed, in some circumstances, provided, of course, that the animal leads a pleasant life, that he/she is killed painlessly, that his/her death does not cause suffering to other animals, and that the killing of this animal allows its replacement by another that would not otherwise have lived. But, Singer adds, if this works well theoretically, the application of this principle is limited. It cannot, normally, justify the killing of wild animals (2011, 111-9), nor may it ever justify factory farming, where animals do not lead pleasant lives, and where the major interest they have in not suffering is sacrificed for the comparatively minor interest humans have in eating their flesh.

It is the question of suffering, and sentience, which truly drives Singer, so that he always goes back to this central idea in his work. He basically claims that the main concern of ethics, or more exactly, utilitarian ethics, is precisely the minimisation of pain and the maximisation of pleasure. He actually moves from an observation on animal physiology and biology to an ethical revolution in the blink of an eye, and he is quite clear about it. For him, there is no need to pin the case for equality to one particular outcome of a scientific investigation. Equality does not depend on intelligence, moral capacity, physical strength, or similar matters of fact; "equality is a moral idea, not an assertion of fact" (1995, 5).

Yet Singer *does* refer to particular scientific data to make his case. He repeatedly stresses the fact that his goal is mainly pragmatic. And as a pragmatic philosopher, he gives instances of animal experiences in his writings. Here is a passage of *Animal Liberation* where he does mention zoological data to advocate a new animal ethic and to shed light on our biased attitude to species:

> While we overlook our own savagery, we exaggerate that of other animals. The notorious wolf for instance, villain of so many folk tales, has been shown by the careful investigations of zoologists in the wild to be a highly social animal, a faithful and affectionate spouse—not just for a season, but for life—a devoted parent, and a loyal member of the pack. […] In keeping with our picture of the world of animals as a bloody scene of combat, we ignore the extent to which other species exhibit complex social life, recognizing and relating to other members of their species as individuals. When human beings marry, we attribute their closeness to each other to love, and we feel keenly for a human being who has lost his or her spouse. When other animals pair for life, we say that it is just instinct that makes them do so, and if a hunter or trapper kills or captures an animal for research or for a zoo, we do not consider that the animal might have a spouse who will suffer from the sudden absence of the dead or captured animal. (1995, 222-3)

Singer does use scientific arguments to explain that some animals are able to make and use tools (an ability that was usually considered to be uniquely human), that some apes have been successfully taught to communicate with American Sign Language. This stands as proof, according to him, that they are self-aware. In 1993, Singer edited a collective volume entitled *The Great Ape Project. Equality Beyond Humanity*, together with Paola Cavalier, to seek an extension of basic rights to chimpanzees, gorillas and orang-utans.[8] The project has since met with some success as several countries, in the Western world, have complied, partly or in full, with its resolutions.[9] In this specific context, the attention he gave to certain animal capacities, such as communicating, or making tools, was clearly strategic and not philosophical. He aimed to question the species divide at its weakest point (the transition between great apes and humans) and to emphasise the possibility for apes to perform tasks that were supposed to be uniquely performed by humans. In so doing, he aimed to extend equality and include not only our own species, but also the species that are our closest relatives—an idea that Elisabeth de Fontenay, as will now be shown, finds particularly offensive.

The Continental Approach: Singer Goes Too Far (Elisabeth de Fontenay)

In the eighteenth century, David Hume observed in his *Treatise of Human Nature* that most systems of moral philosophy make, at some point, an unperceived transition from premises whose parts are linked by "is" to conclusions whose parts are linked by "ought," thus expressing a new relation, which seemed impossible to Hume. His observation later became known as Hume's Law and claims that one cannot reach a normative, or morally binding statement from a premise that does not itself contain at least one normative element. In plain words, rights should not be based on facts.

Elisabeth de Fontenay, a contemporary French philosopher who has examined the relationships between humans and animals from a continental philosophical perspective[10] in books such as *Le Silence des Bêtes*, has conjured up Hume's Law in her review of Singer. In an article published in *Le Débat*, she gets to grips with Singer's propositions (2000). She strongly rejects Singer's use of ethology, biology and zoology as normative authorities and reaffirms human exceptionalism in the following terms:

> Zoologists, paleoanthropologists and ethologists are researchers. We are certainly under no obligation to yield, without scrutiny, to the ethical and political interpretations they suggest, sometimes naively, from their results. For their work does not offer them enough resources to free us from the attachment to the specific attributes of mankind which—from Genesis, Exodus, Artistotle's *Politics*, to the Declaration of the Rights of Man—

> have been the most consistent approach to our use of reason. (2000, 141)

De Fontenay is right to emphasise Singer's logical leap, which works like this:

- First premise. Animals are sentient [this is a biological, descriptive fact].
- Second premise. We ought to consider the interests of all sentient beings [this a normative statement].
- Conclusion. Ethics requires us to consider the interests of animals equally [this is a normative, deontic conclusion].

The hidden transition that de Fontenay finds particularly abhorrent lies in the marginal cases argument which Singer so powerfully invokes in most of his ethical writings on animals, and especially in *Practical Ethics*. Through this argument, Singer attempts to demonstrate that if animals do not have direct moral status, then neither do such human beings as infants, the senile, the severely cognitively disabled, and other "marginal cases" of humanity.

We see humans as special beings, Singer contends, and warrant them special moral status accordingly, because they are, as a species, rational, autonomous, able to speak, self-aware, able to see themselves in the future, capable of making choices, etc. We consider these characteristics critical and deny animals direct moral status because they do not possess these properties. Yet, our ethical system has also, traditionally, considered the interests of "marginal humans" (the comatose, the newborn, those with severe mental disabilities), who do not possess such capacities, in the same way as it has considered the interests of other humans. This is inconsistent, Singer claims. Why should the interests of marginal humans count more than the interests of animals with similar or even greater capacities? The conclusion of the marginal cases argument, according to Singer, is that we ought to grant no less importance to the interests of animals than we do to the interests of humans generally.

Thus, Singer exposes anthropocentrism to a dilemma. If the criterion for moral standing is pitched high enough to exclude animals, it will also exclude some human beings; but if it is pitched low enough to include all human beings, it will also include a large group of animals. Is it a levelling up of animals which are brought into the moral world of humans or rather a levelling down of newborn children, patients in a deep coma, the severely handicapped? Singer hopes that he will be interpreted correctly. He does indeed advocate a levelling up, or rather, a sort of middle position, as explained as early as 1975 in the first edition of *Animal Liberation*:

> What we need is some middle position which would avoid speciesism but would not make the lives of the retarded and senile as cheap as the lives of pigs and dogs are now, nor make the lives of pigs and dogs so sacrosanct that we think it wrong to put them out of hopeless misery. What we must do is bring nonhuman animals within our sphere of moral concern and cease to treat their lives as expendable for whatever trivial purposes we may have. (1995, 20)

However, de Fontenay opposes his conclusion. There is one property that is immune to this line of argument, the property of being human. She replies that rights should not be proclaimed from facts, be they biological or other; rather, rights should be proclaimed by a State from a metaphysical, transcendent and immanent conception of human natural rights, or they should be invented, instituted (2000, 141); but what, one may respond, may this metaphysical conception be based on? This line of reasoning seems also based on a leap of logic that continental philosophy tries to fill in as best it can. De Fontenay claims that one should start with the metaphysics of liberty, which another French contemporary philosopher, Luc Ferry, concurs with. For Ferry, a moral being is first and foremost defined by "his liberty, conceived as a faculty to force himself out of the idiosyncratic, natural and historical determinisms which weigh heavy on all natural (premoral) life forms." (164)

Like de Fontenay, Ferry also rejects the marginal cases argument. He thinks newborn children, comatose patients and the severely disabled will one day cease to be such marginal humans, or might cease to be so, or might not have become such passive citizens, whereas a chimpanzee will always remain a chimpanzee, precluding any kind of ethical reciprocity. For Ferry, it seems impossible to ignore this dichotomy.

Nevertheless, in *The Great Ape Project*, Singer and Cavalieri hold that scientific progress has definitely undermined human exceptionalism, which de Fontenay finds unacceptable. She believes that only a patient and careful deconstruction of humanism may encourage people to respect animals without offending mankind (2000, 151). However, I would also like to emphasise that continental ethic has itself bridged the gap between descriptive premises and normative statements by granting humans a special moral status based on their being endowed with special characteristics (autonomy, language, a capacity to see oneself in the future, etc.).

As any philosopher would be, Singer is aware of Hume's doctrine. He tackled the issue of the opposition between "is" and "ought" in one of his very first articles published in 1973. But he only broached the subject to set it rapidly aside. In his view, it is essential, in any given work on ethics, to clarify in what sense one is using terms like "moral judgment," and what follows from the use of such a term. But this should just be a preliminary. It should not be the subject-matter of so many works of moral philosophy published in the English-speaking world in the late twentieth century.[11] This question is, in his view, mainly a question of terminology which philosophers should not lose too much time on. Instead they should concentrate, as Singer does, on urgent ethical questions and on how statements of facts are connected practically with reasons for acting and not, theoretically, with moral judgments in general. Once again, Singer shows how pragmatic his approach is:

> The problem of how statements of fact are related to moral judgments has dominated recent moral philosophy. Associated with this problem is another, which has also been given considerable attention—the question of how morality is to be defined. The two issues are linked, since some definitions of morality allow us to move from statements of fact to moral judgments, while others do not. In this article I shall take the two issues together, and try to show that they do not merit the amount of attention they have been given.
>
> I hope it is now clear that the issue that really matters, that is of practical significance, is how statements of fact are connected with reasons for acting, and not how statements of fact are connected with moral judgments. (1973, 51)

For Singer, facts matter; his job is not just about philosophy and critical reasoning. It is quite revealing that his groundbreaking work *Animal Liberation* is filled with practical suggestions for how we are to live and how we should change our habits. But while these suggestions may seem too radical for continental academics like Elisabeth de Fontenay or Luc Ferry, others, like Holmes Rolston III, believe on the contrary that they do not go far enough.

The Biocentric Approach: Singer Does not Go Far Enough (Holmes Rolston III)

In a paper entitled "Respect for Life: Counting what Singer Finds of no Account," American theologian and philosopher Holmes Rolston III discusses the principle of equal consideration of interests. Within a broad framework of environmental ethics combined with a theological perspective, Rolston challenges sentience as the most morally significant concept, and promotes instead the notions of value and species.

While Singer goes beyond the anthropocentric position, as advocated by de Fontenay, and embraces a pathocentric approach in which sentience is the morally relevant property, Rolston goes even further and advocates a biocentric position, extending the scope of human responsibility to all living things, as exemplified in this figure:

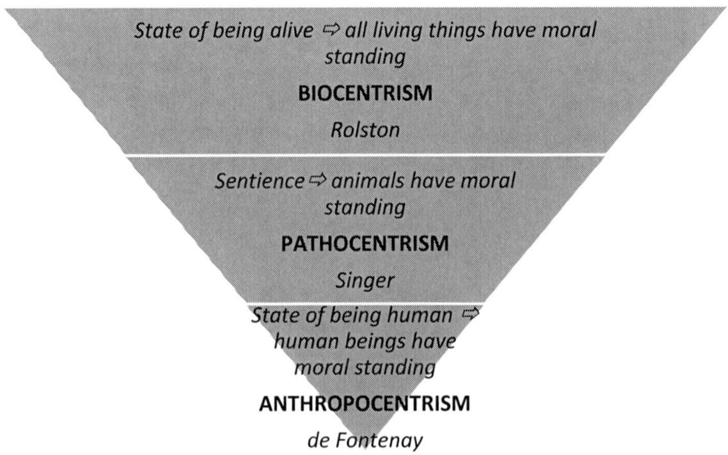

Adapted from Martin Gorke *The Death of our Planet's Species. A Challenge to Ecology and Ethics,* Washington: Island Press, 2003 figure 2, p. 122.

Rolston endorses the central tenet of biocentrism according to which each living being has a good of its own and having a good of its own is the ground of a being's intrinsic value, whether this being is sentient or not, whether it is an animal, a plant, or a micro-organism.[12] As propounded by Paul Taylor in a seminal book entitled *Respect for Nature: A Theory of Environmental Ethics* (1986), biocentrism suggests that we have a *prima facie* moral duty to preserve or promote the good of all living things as ends in themselves. Hence, each organism has a purpose and reason for being, with no superior category of individuals. In Taylor's theory, the community of living organisms is interdependent, all are members, along with other species, and on equal terms.

As a proponent of biocentrism, Rolston regrets Singer's apparent disregard for environmental ethics and the attempt to value life at all its ranges and levels. The environmentalist would like to see the animal liberationist care for a biospheric Earth globally, more so than just for some if its inhabitants, considered individually. Rolston deplores the fact that insentient life seems to be of no account in Singer's animal liberation ethic and that plants, for instance, leave Singer "at a loss" (1999, 247). He further contends that Singer's argument works only for vertebrates who form a tiny minority of living things and that Singer has yet to count the vital processes such as speciation, natural selection, and ecosystemic communities in which these organisms live. For Rolston, such an approach to ethics is mistaken. It is necessary to locate the mistake, or Singer's potential victory for vertebrates would "prove pyrrhic" (1999, 247) in the fight for a broader green ethic. Rolston's concern, as an environmentalist, is whether plants and other insentient beings can be loci of value. He suggests that, if things do not matter *to* plants (in Singer's utilitarian ethics, they cannot be said to have any "preference" whatsoever), things matter *for* plants. He claims that organisms are not moral systems, for there are no moral agents in nature. Rather, Rolston explains, organisms are axiological systems. So the plant comes to existence, grows, reproduces, and tries to live as long as possible. The physical state that the organism defends is a valued state. Organisms may have no preferences, but they do have their own standards. So that, according to Rolston, each organism has a *good-of-its-kind*. In the same way, a species has a value in itself, and should be protected as such. This kind of value falls outside the individualistic and pathocentric framework advocated by Singer. As Rolston makes his case:

> The species is a bigger event than the individual, regardless of whether the member individual has interests or sentience. Much of what we have said about individual organisms as non-moral normative systems can be resaid, *mutatis mutandis*, of species. The species line is the vital living system, the whole, of which individual organisms are the essential parts. The species line is the vital living system, the whole, of which individual organisms are the essential parts. Processes of value that we earlier located in an organic individual reappear at the specific level: defending a particular form of life; pursuing a pathway through the world, resisting death (extinction), regeneration maintaining a normative identity over time, creative resilience discovering survival skills. The species has a good-of-its-kind. Situations can be better or worse for species. (1999, 261)

For Rolston, the individual is subordinated to the species, not the other way round. Since the loss of a species puts an end to genetic possibilities, and the deliberate destruction of a species would show disrespect for the very biological processes which make possible the birth of individual living things, species are more valuable than individual specimens (Brennan, and Lo, 2011). His focus is on what he calls "the vital living system"; it is respect for nature and, in particular, for natural species. Singer is biased when he claims that only the conscious experiences of the subjective lives of vertebrates have value (Rolston 1999, 265-7). This means that only a late product of the system, psychological life, is valued. This is too narrow an entry point for Rolston who suggests instead that one should value life, as a whole, arising as a creative process on Earth. This is, he challenges, a more promising and a more comprehensive starting point than the one based on the pains, pleasures and preferences of a small part of its occupants. Indeed, in a manner which finds a clear echo in John Baird Callicott's recent *Thinking like a Planet. The Land Ethic and the Earth Ethic*, Rolston sees the planet as the most relevant entity of all, because it is the only entity capable of producing Earthbound values.

The question of valuing individuals or species is, according to Rolston, partly a scientific and partly a philosophical problem (this echoes Singer's presentation of the question of animal pain as both scientific and philosophical):

> Can a species be value-able all by itself, able to value at the species-level? A species has no self. There is no analogue to the nervous hook-ups or circulatory flows that characterize the organism. But now we must ask whether singular identity conserved is the only process that is valuable. A species is another level of biological identity reasserted genetically over time. Identity need not attach solely to the centered or modular organism; it can persist as a discrete pattern over time. (1994, 19-20)

He further contends that animals bear in themselves the value of their species; they are all, individually, meant to die, and they assume reproduction so that their *species* can survive, not them, individually. Their organs are "defending the line of life bigger than the somatic individual" (1994, 19-20).

Finally, Rolston offers a wider approach to animal ethics than Singer. He expresses it in terms of value and species, instead of sentience, but Singer has something to say for his defence. He sees a major flaw in Rolston's system: if a clash occurs between the survival of a species or of an individual organism, the species takes precedence, because it is a bigger event. If so, Singer writes back, why stop at rabbits or goats? If an individual human being is a threat to a plant species, should he/she be removed?

> it is quite misleading to summarize my view by saying that I insist 'that "there is nothing to be taken into account" in all the display of vitality in insentient life'. Such statements ignore—strangely, for an environmentalist—the aesthetic, scientific and recreational values of preserving natural ecosystems. For the same reasons, I do not have to agree that 'all is well and good' when a giant tree is cut down each year to serve

> as a national Christmas tree on the White House lawn. Rolston is just plain wrong when he says that I have 'no grounds on which to recommend' his idea of leaving a tree in place, decorating it and telling people why it was left uncut. I would recommend that this be done to symbolize the importance of preserving our beautiful and ancient forests; which provide homes and food for many wild animals and help to keep the climate of our planet stable. I would also use the opportunity to tell people about the waste of resources involved in the old practice of felling and transporting a large tree across the country. Instead of seeing nature as something for us to exploit and destroy to satisfy quite trivial desires of our own, I would say, we should see it as a precious heritage that we should preserve in order to pass on to our children and grandchildren. (Jamieson, 332)

Conclusion

In 2010, Peter Singer participated in a debate held at Arizona State University with other preeminent thinkers in philosophy, physics, and psychology, including Sam Harris, Steven Pinker and Simon Blackburn, on the question of whether science can tell us right from wrong. His contribution was basically that, although science does provide us with important material regarding what is ethical and what is not, it does not provide us with moral premises. That is to say, it cannot provide us with the basic values that we have to work with.

In many cases, science can give indications as to how we can better implement the values we have, and on which we agree. But it cannot dictate these values to us. Therefore, Singer rejects the use of biology or ethology as normative authorities. We ought to be concerned about the welfare of animals, he claims, because they are sentient, and science tells us so. Similarly, we ought to be concerned about the preservation of species because they participate in the ecological balance, as science has shown. On the other hand, Singer adds, science cannot tell us how many animals we can use in a scientific experiment to improve medical knowledge, for instance. This is a question that is open to discussion.

More importantly, Singer argues, the value we have, on which we agree, and which serves as a moral premise, is that pleasure is a good that ought to be promoted, and pain, as the negation of pleasure, ought to be minimized.[13] Since most of us readily accept that we should keep to a minimum the pain that we might cause others, Singer believes that once science has shown that animals can suffer, and we recognise it, it is only logical to include them in our moral calculations.

In his advocacy of animal liberation, Peter Singer refers to scientific notions. He also openly shows, as Rolston does, that his concern has to do, at certain points, both with science and philosophy. But for all that, his final goal is neither philosophical nor scientific. It is practical. He does not recoil from this fact, he is actually ready to admit that his work is radical and ambitious (he wants to change the world), and based on facts (scientific facts, but also statistical material, or images of the actual situation of animals in agribusiness and research, etc.). In Singer's work, science is just a tool among others, or an ally, in an overall strategy to build a new animal ethic. In this respect, his ideas have partly been echoed in the Western world, where the welfare of animals raised for food is now a subject of concern for politicians and legislators—certainly more so than fifty years ago. Standard laying hen cages are now banned in the European Union (2012) and in California (2015), while veal calf stalls have been banned in the UK since 1990 and in the European Union since 2007. The most striking change illustrating the growing acceptance of animal ethics, though, is perhaps the recent historic court hearing to determine whether two chimpanzees could obtain writs of habeas corpus for being "unlawfully detained" at Stony Brook University.[14]

If an American court agrees to hear a case on the legal personhood of chimpanzees, then animal interests have been more firmly put onto the political and legal agenda, and Singer's principle of equal consideration has indeed made some headway.

Bibliography

Aristotle, *Politics*. 1916. London: J.M. Dent and Sons.

Arneson, Richard J.1999. "What, if Anything, Renders All Humans Morally Equal?" in Dale Jamieson (ed.). *Singer and His Critics*, Oxford: Blackwell, pp. 103-128.

Bentham, Jeremy. 2007. *An Introduction to the Principles of Morals and Legislation*. New York: Dover. [1780].

Brennan, Andrew and Lo, Yeuk-Sze. 2011. "Environmental Ethics," *The Stanford Encyclopedia of Philosophy,* Fall Edition. Edward N. Zalta (ed.). To be consulted at: http://plato.stanford.edu/archives/fall2011/entries/ethics-environmental/. (accessed 9 December 2014)

Callicott, J. Baird. 2014. *Thinking like a Planet. The Land Ethic and the Earth Ethic*. New York: Oxford University Press.

Cavalieri, Paola and Peter Singer (ed.).1996. *The Great Ape Project. Equality Beyond Humanity.* New York: St Martin.

Cyrulnik, Boris, Elisabeth de Fontenay, Peter Singer. 2013. *Les Animaux aussi ont des droits*. Paris: Editions du Seuil.

Dardenne, Emilie. 2008a. "Les nouvelles frontières de l'altérité ? Peter Singer contre le spécisme," *L'autre*, Janine Dove-Rume, Michel Naumann, Tri Tran (eds.). Tours : Université François-Rabelais, pp. 469-488.

—. 2010a. "The Reception of Peter Singer's Theories in France". *Society and Animals. Journal of Human-Animal Studies* 18, vol. 2, pp. 205-218.

—. 2010b. "From Jeremy Bentham to Peter Singer," *Revue d'Etudes Benthamiennes* n°7, septembre, To be consulted at: http://etudes-benthamiennes.revues.org/204. Followed by a reply by Peter Singer: http://etudes-benthamiennes.revues.org/212.

Descartes, René. 1646. "Lettre au Marquis de Newcastle," 23 novembre 1646, *Oeuvres et lettres*. Paris: Gallimard, bibliothèque de la Pléiade, 1983, b, pp.1256-1257.

Donaldson, Sue and Will Kymlicka. 2011. *Zoopolis. A Political Theory of Animal Rights*. Oxford: Oxford University Press.

Ferry, Luc. 2000. "Des 'droits de l'homme' pour les grands singes? Non, mais des devoirs envers eux, sans nul doute". *Le Débat*, 108, février, pp. 163-167.

Fontenay, Elisabeth de. 2000. "Pourquoi les animaux n'auraient-ils pas droit à un droit des animaux ?" *Le Débat*, 108, février, pp. 138-155.

Gorke, Martin. 2003. *The Death of our Planet's Species. A Challenge to Ecology and Ethics*. Washington: Island Press.

Holton, Richard and Rae Langton. 1999. "Empathy and Animal Ethics," in Dale Jamieson (ed.), *Singer and His Critics*. Oxford: Blackwell, pp. 209-232.

Hume, David. 2007. *A Treatise of Human Nature*, David Fate Norton and Mary J. Norton (eds.), 2 vols. Oxford: Oxford University Press. [1739]

Jamieson, Dale (ed.). 1999a. *Singer and His Critics*. Oxford: Blackwell.

—. 1999b. "Singer and the Practical Ethics Movement," in Dale Jamieson (ed.), *Singer and His Critics*. Oxford: Blackwell, 1999, pp. 1-17.

Krantz, Susan Lufkin. 2002. *Refuting Peter Singer's Ethical Theory. The Importance of Human Dignity*. Westport: Praeger.

Rolston III, Holmes. 1994. "Value in Nature and the Nature of Value," in *Philosophy and the Natural Environment*. Robin Attfield, Andrew Belsey (eds.). Cambridge: Cambridge University Press, pp. 13-30.

—. 1999. "Respect for Life: Counting what Singer Finds of no Account," in Dale Jamieson (ed.), *Singer and His Critics*. Oxford: Blackwell, pp. 247-268.

Singer, Peter. 1973. "The Triviality of the Debate Over 'Is-Ought' and the Definition of 'Moral'". *American Philosophical Quarterly*, X, 1, January, pp. 51-56.

—. 1995. *Animal Liberation, Second Edition with a New Preface by the Author*. London: Pimlico [First Edition 1975, Second Edition Jonathan Cape, 1990].

—. 2011. *Practical Ethics, Third Edition*. Cambridge: Cambridge University Press [First Edition, 1980, Second Edition, 1993].

—. 1999. "A Response," in Dale Jamieson (ed.), *Singer and His Critics*, Oxford: Blackwell, pp. 269-335.

—. 2010, "The Great Debate—Can Science Tell Us Right from Wrong?," talk given at Arizona State University on November 6. To be consulted at: http://thesciencenetwork.org/programs/the-great-debate/peter-singer. Accessed 21 October 2014.

Speedy, Andrew. 2003. "Global Production and Consumption of Animal Source Foods," *The Journal of Nutrition*, November 1, vol. 133, n° 11, p. 4048S

Notes

[1] Between 1960 and 2000, the production of meat was multiplied by two in developed countries, and by three in developing countries, reaching around 220 million metric tons in 2000. See Andrew Speedy, "Global Production and Consumption of Animal Source Foods," *The Journal of Nutrition*, November 1, 2003, vol. 133, n° 11, p. 4048S.

[2] This is illustrated for instance in the Great Ape Project launched by Peter Singer and Paola Cavalieri in 1993 to demand the extension of the community of equals to include all human and nonhuman great apes. The book edited by Singer and Cavalieri includes contributions by ethologists and primatologists (Jane Goodall, Richard Dawkins), anthropologists (Jared Diamond, R.I.M. Dunbar), as well as ethicists (Peter Singer, Paola Cavalieri, Tom Regan, Steve Sapontzis, James Rachels) and psychologists (Francine Patterson, Robert W. Mitchell).

[3] Aristotle claimed that animals rank far below human beings in the Great Chain of Beings and that: "Plants exist for the sake of animals, and brute beasts for the sake of man—domestic animals for his use and food, wild one (or at any rate most of them) for food and other accessories of life, such as clothing and various tools. Since nature makes nothing purposeless or in vain, it is undeniably true that she has made all animals for the sake of man." *Politics*, London: J.M. Dent and Sons, 1916, p. 16. Later, Descartes wrote: "[…] si elles pensaient ainsi que nous, [les bêtes] auraient une âme immortelle aussi bien que nous ; ce qui n'est pas vraisemblable, à cause qu'il n'y a point de raison pour le croire de quelques animaux, sans le croire de tous, et qu'il y en a plusieurs trop imparfaits pour pouvoir croire cela d'eux, comme sont les huîtres, les éponges, etc. " "Lettre au

Marquis de Newcastle," 23 novembre 1646, *Oeuvres et lettres*, Paris: Gallimard, bibliothèque de la Pléiade, 1983, b, pp.1256-1257.

[4] The founder of utilitarianism, Jeremy Bentham, argued as early as 1780 that sentience should be the only criterion used to establish the limits of our moral sphere: "What [...] is it that should trace the insuperable line? Is it the faculty of reason, or, perhaps, the faculty of discourse? But a full-grown horse or dog is beyond comparison a more rational, as well as a more conversable animal, than an infant of a day, or a week, or even a month, old. But suppose the case were otherwise, what would it avail? the question is not, Can they *reason*? nor, Can they *talk*? but, Can they *suffer*?," *An Introduction to the Principles of Morals and Legislation*, New York: Dover, 2007 [1780], p. 311 n.

[5] Here is Singer's explanation for this argument: "The nervous systems of animals evolved as our own did, and in fact the evolutionary history of human beings and other animals, especially mammals, did not diverge until the central features of our nervous systems were already in existence. A capacity to feel pain obviously enhances a species' prospects for survival, since it causes members of the species to avoid sources of injury". *Animal Liberation*, New York: Avon Books [2nd edition], 1990, p. 10-12, 14-15.

[6] *Ibid.*

[7] Singer quotes Lord Brain, "Presidential Address," in C.A. Keele and R. Smith, eds., *The Assessment of Pain in Men and Animals*, London: Universities Federation for Animal Welfare, 1962.

[8] According to Singer and Cavalieri, here are the basic principles that should govern our relations with each other in a community of equals that includes humans and great apes:

"1. The Right to Life.

The lives of members of the community of equals are to be protected. Members of the community of equals may not be killed except in very strictly defined circumstances, for example, self-defence.

2. The Protection of Individual Liberty

Members of the community of equals are not to be arbitrarily deprived of their liberty; if they should be imprisoned without due legal process, they have the right to immediate release. The detention of those who have not been convicted of any crime, or of those who are not criminally liable, should be allowed only where it can be shown to be for their own good, or necessary to protect the public from a member of the community who would clearly be a danger to others if at liberty. In such cases, members of the community of equals must have the right to appeal, either directly or, if they lack the relevant capacity, through an advocate, to a judicial tribunal.

3. The Prohibition of Torture

The deliberate infliction of severe pain on a member of the community of equals, either wantonly or for an alleged benefit to others, is regarded as torture, and is wrong."

Paola Cavalieri and Peter Singer (Ed.), *The Great Ape Project*, *op. cit.*, p. 4.

[9] In 1999, in New Zealand, after scientists and lawyers petitioned the parliament to pass a bill conferring "rights" on apes, the government granted them legal protection from animal experimentation. A few years later, the Spanish parliament passed a resolution that gave apes the right to life and freedom (2008). In 2010, after months of heated debates, the European Union made the use of apes in animal testing illegal. Under the new legislation, experiments on chimpanzees, gorillas and orangutans are to be banned and restrictions set on the use of primates in general.

[10] The phrase "continental philosophy" was first used in the second half of the twentieth century by Anglo-American and Australian philosophers to refer to thinkers influenced by European political philosophy. It is opposed to the analytical approach.

[11] Peter Singer, "The Triviality of the Debate Over 'Is-Ought' and the Definition of 'Moral,'" *American Philosophical Quarterly*, X, 1, January 1973, p. 51-56.

[12] Andrew Brennan and Yeuk-Sze Lo, "Environmental Ethics," *The Stanford Encyclopedia of Philosophy,* Fall 2011 Edition, Edward N. Zalta (ed.), http://plato.stanford.edu/archives/fall2011/entries/ethics-environmental/ Viewed 9 December 2014.

[13] This is the utilitarian principle, which is actually not universally accepted as a starting point for any ethical system, and which is what raises opposition from de Fontenay.

[14] This hearing took place in May 2015 in New York City. The lawyers representing the two apes argued before Manhattan Supreme Court Justice Barbara Jaffe that the animals were unlawfully imprisoned and should be relieved. The question at stake was how to interpret the legal principles of unlawful detainment and the right to bodily liberty for nonhuman primates; and how to classify them, knowing that, as the two lawyers from the Nonhuman Rights Project argued, they are autonomous and self-determining beings. At the time of writing, Judge Jaffe had not yet ruled on the case.

PART III:

THE RHETORICAL POWER OF SCIENCE

CONSTRUCTING WEAKNESS: SCIENTIFIC RHETORIC AND VICTORIAN SENSIBILITIES

JULIE HOMCHICK

As the pursuit of scientific knowledge gained momentum in 19th century England, so did the dissemination of scientific information to the public. Gaining authority as an empirical, irrefutable source of truth, science embodied the unquestioned power of fact that could be used to rationalize practices, underscore ideological thinking and manipulate the public. Through the rhetoric and authority of science during that time, scientific texts could conceal ideological leanings while fabricating a scientifically-informed consciousness among the public. One dominant example of the concealment of ideological dispositions is the reinforcement of the construction of women's multi-faceted inferiority within scientific texts. Human sciences of the day, such as phrenology, eugenics and evolution, relied on the authority of factual, scientific language while simultaneously playing a role in the reification of Victorian gender norms particularly by underscoring the traits of female inferiority and domesticity. The study below details the ways in which practitioners of these sciences—including prominent figures such as Franz Gall, Francis Galton, Charles Darwin and others—advanced scientific proposals that reinforced the "separate spheres" of gender identified by Leonore Davidoff and Catherine Hall in Victorian society.

Background

The public discourse of science in Victorian England largely stemmed from the advances of Newtonian physics, which became a near obsession for Western philosophers during the century prior. Isaac Newton's physical laws and Francis Bacon's theories of methodology created a scientific English enterprise that continued through the British Industrial Revolution (Gay 174). With the establishment of the British Association for the Advancement of Science in 1831, England continued its quest for empirical knowledge. Aside from solidifying a scientific community for scientists to work within, the organization also aspired to promote accessible scientific discourse in the public sphere through journals, newspapers and lectures. The public's growing interest in science emerged largely as an effect of the utilitarian values of the period with which the public began to consider scientific studies as merely useful in yielding isolated facts, ignoring contextual information regarding methodology, selection of evidence and differences between causation and correlation. Richard Yeo summarizes this problem well:

> […] the attractiveness of easily acquired information had taken precedence over the careful study of scientific reasoning, and profound theoretical speculation had been overshadowed by the demand for immediately useful knowledge. Popularization, by catering to these tastes, had encouraged a perception of natural knowledge as a collection of interesting and useful facts, and this view militated against a proper appreciation of science. (177)

In this respect, the benefits and uses of science minimized the nuances of experiments and scientific research. This resulted in the public consuming commodified, practical scientific facts and devices, rather than understanding the nature of scientific theory and inquiry. In addition to public education events held by the British Association and writings and reference books for the public like W.T. Brande's 1842 *A Dictionary of Science, Literature and Art,* celebratory spectacles of science, like the Great Exhibition of 1851, dazzled the public with scientific achievements from printing presses to cigarette machines. This commodification was even reinforced through the theories of some scientists themselves, such as eugenicist Karl Pearson, who advocated science's infallibility and endless applicability: "The field of science is unlimited; its solid contents are endless, every group of natural phenomena, every phase of social life, every stage of past or present development is material for science" (Pearson, *Grammar of Science,* 15). With Pearson's view, a 'scientification' of everything was possible.

However, some scientists grew uncertain regarding the public's reception of scientific information, which raised questions concerning nonscientists' ability to "distinguish between fact and hypothesis…" (Yeo, 173). British Association member Clerk Maxwell, in his 1871 "Introductory Lecture on Experimental Analysis," charged: "Such indeed is the respect paid to science, that the most absurd opinions may become current, provided they are expressed in language, the sound of which recalls some well-known scientific phrase" (Maxwell, 242). Maxwell identified the danger of the authority of scientific language, seeing it as encouraging the public's acceptance of any supposed scientific information. Sir George Cornewall Lewis also saw the potential detriments of scientific language: "No species of imposture is so captivating, so well-suited to the present time, and consequently so likely to meet with temporary success, as that which assumes the garb, and mimics the phraseology, of science…" (55). Both Maxwell and Lewis's remarks specifically point to the power of scientific language—according to them, such terminology and phrases carried such significant import with the public that the actual contents of a scientific claim could be unfounded and meaningless.

But even with existing skepticism as to the power of scientific language among some scientists, such specialist and fact-driven language maintained its authority in the public sphere, creating what Neil Postman refers to as "Scientism," or an unquestioned faith in the ultimate authority of science (Postman, 144-163). This faith Postman identifies is rooted in Peter Allan Dale's identification of a Victorian "religion of science" where a faith in science began to shake the ground of metaphysical romanticism (7). Dale goes on to claim this religion of science as the most powerful, authoritative construction of a belief system: "No religious belief, no form of metaphysics, no social philosophy so powerfully affected the thought of educated men and women as this new religion of science" (11). The faith in science could ultimately construct science's undisputed supremacy and power as a purveyor of objective information, paradoxically rendering its content further subject to hidden biases.

While science became increasingly popularized in the mid-1800s, the English working-class began demanding more rights as part of the Chartist Movement so that basic liberties could be granted to all people (men and women) regardless of property-owning status. As Anna Clark points out, though, this movement ultimately reinforced traditional gender divides through a defense of the family, where women were viewed primarily as wives and mothers and where "domesticity was a privilege all, not just the middle class, should enjoy" (193). This relates directly to Davidoff and Hall's work on the "separate spheres" of nineteenth century England. In their book *Family Fortunes*, the authors show how industrialization in the late-1700s gradually strengthened the divide between work and home where men participated in the public sphere while women stayed home in the private, domestic sphere. The authors argue that by the 1840s "the expectation was that middle-class women should not be gainfully employed and many were able to retreat to a domesticated life in their suburban villas and gardens" (xiv). And science did not exist independently from this middle-class discourse on separate spheres. As

Yeo suggests, "the popularity of science was symptomatic of the material values of the new middle class" (76) and the popularity of such texts in Victorian society indicates their discursive ability in shaping beliefs and attitudes about social and cultural issues, including gender norms. In her work on the construction of female inferiority in 19th century Western Europe (specifically France), Joan Wallach Scott shows how authoritative documents—including scientific texts—acted as "a participant in the production of knowledge that legitimized the exclusion or subordination of women" (Scott, 26). In a time where "facts and figures served as a kind of self-evident truth" (114), she shows how the construction of female inferiority through a "natural" dependency (129) on males operated "under the guise of objectivity" (118). By attending to similar fact-driven language in the texts analyzed below, this work illustrates how scientific language specifically functioned to maintain the "separate spheres" identified by Davidoff and Hall. In texts on phrenology, eugenics, and human evolution—sciences that focused on human traits and behaviors—Victorian gender norms of female domesticity imbued scientific thinking in ways that reinforced and strengthened the public's perception of gender.

Phrenology, Femininity and the Fragile Intellect

The practice of phrenology, while largely developed in the late eighteenth century by Viennese doctor Franz Joseph Gall, gained its popularity during the early and mid-1800s in England. Advocates of this originally Aristotelian proposal (Cooter 4) used scientific terms and measurements to illustrate how character and dispositions of individuals were represented through the indentions and irregularities of their skulls. According to phrenology's advocates, Gall's methods could be used to determine an individual's class, intelligence, benevolence or strength in character and was often a tool in Victorian portraiture and fictional character development in works by George Eliot and other authors (Witemeyer, 45). As evidenced in their writings, authors like Gall and his colleagues attributed the characteristics of weakness, benevolence and domesticity to women, while the attributes of intelligence, strength and confidence were found on the skulls of men. Given phrenology's scientific standing during the nineteenth century, these texts embodied such gender norms within seemingly neutral scientific language. Roger Cooter supports the importance of attending to the ideological implications of such language:

> Because phrenology now appears in many respects as a Dickensian

> caricature of science, its study can also magnify both the way in which social and ideological meanings are implicit in science and the way in which these meanings are redefined according to social contexts. Received in Britain as a scientific system of exaggerated symbols, phrenology demonstrates how these symbols may serve as alternative codified renderings of social structures and relations (i.e., as 'mediations'). (9)

By unpacking the language used by notable phrenologists, it becomes apparent that the domesticity, intellectual inferiority, and weakness of women were prescribed through readings of the skull, ultimately showing how the language of science served as an aid in the construction of women's domestic roles in 19th century England.

Gall was reticent to publish any of his work until he felt that he had properly developed his ideas, but phrenology nonetheless found its way to the British public through reports of his lectures and the writings and touring lectures of popularizers J.G. Spurzheim (one of Gall's students) and George Combe. It soon became "enormously successful in Britain," as Trevor Levere points out, and "in 1832 there were approximately thirty British phrenological societies" (181). Reinforcing its popularity, figures such as Queen Victoria, Prince Albert, George Eliot and "a random catalogue of 19th century figures" accepted and practiced phrenology in both work and life (Davies, 39). As Cooter observes, phrenology "appeared to be argued by facts alone, to which were added sensible-sounding practical conclusions," and it was particularly persuasive through the presence of "the very artifacts of medical science—the skulls, busts, casts, and charts of brains with which lecturers surrounded themselves" (75). In the case of phrenology, the public could easily observe practitioners and understand its implications in their own lives, making it both more believable and ideologically powerful.

While Gall spoke regularly to the public on the topic of phrenology, many of his lectures were not transcribed directly but were reported on by viewers. In a series of lectures from 1805 given in Dresden, Germany, Charles Blöde recounts the message of Gall for other scientists and members of the public. Given that Gall's major work, *On the Function of the Brain*, was not translated into English until 1835, texts like Blöde's offer insight into the messages public audiences were receiving about Gall and phrenology. Even if Blöde's account is not entirely accurate, it still functions as a primary source of insight into how the public was presented with Gall's work.

Unpacking Gall's language about the human skull, it becomes evident how he used the authority of science to explain and reinscribe notions of proper female roles and traits in contrast to those of men. In Blöde's account of these lectures, Gall first does so by indicating that a parental inclination is more pronounced in the female sex, insofar as the fondness for offspring is found to be "always more apparent in the female than in the male." He explains this by locating the trait through physiological means: "In the human species it gives the female skull a peculiar pointed form behind, and the sex of a child can be known by the single criterion" (Blöde, 72). By restricting the fondness for offspring to females, Gall fortifies existing notions of maternity, domesticity and child-rearing. He further reinforces this by emphasizing the absence of this physiological trait in men, suggesting the male's position to be outside of the home and in the public, professional sphere. His claims about the "organ of locality" serve a similar purpose—when this organ is present, it indicates an inclination to all the sciences and arts, a fondness for traveling and well-developed spatial skills. Gall's study of the skull shows this organ to be well-developed in the males of most species and "even in man developed to a much greater degree in the male than in the female sex" (75). This again reinforces the belief that men are suited to remain in the public, professional sphere through their better understanding of every field of study and their inclination to travel. Since the female sex's physiology shows an underdeveloped organ of locality, Gall implies women's ability in such public arenas is inadequate, again confining them to the domestic world. Underscoring their propensities to thrive in the domestic sphere, women also housed a well-developed organ of vanity, which Gall claimed was "found more often distinctly expressed in women than in men" (82). This was apparent through "two hemispheric protuberancies," that offered physiological proof of the trait (82). In these passages that Blöde notes, women are suggested to have naturally and anatomically developed as domestic and idle individuals who had time for adornment of their homes and selves.

Beyond early accounts of Gall's work by individuals like Blöde, Gall's published works, most notably *On the Function of the Brain*, continued to disseminate similar divisions of inclinations and abilities found between the two sexes. In this book, Gall first explains some of the different physiological structures that distinguish the skulls of men and women. Women's brains, he argues, are smaller toward the anterior-superior part of the forehead, giving them small and short foreheads in comparison to men. Their cerebellums are often smaller as well whereas the parts of the brain in the superior region are larger. While perhaps benign on first glance, these findings were then interpreted by Gall: "Now these differences explain perfectly, the superiority of the intellectual faculties in man, and the greater energy of the love of children in women, &c." (182). Here Gall returns to earlier suggestions that the division of abilities in men and women are scientifically supported. The shapes of skulls, he argues, directly indicate why men have domain over the public sphere with their intellect, while women remain domestic at home with children. He goes on to claim "that it is not education, but nature, which, by means of a varied organization, has assigned to each sex its particular sphere of moral and intellectual activity" (183). An authoritative force, nature is identified as the deterministic cause of these divisions of abilities and ultimately of the divided spheres of public and private life.

As the ideas of Gall gained popularity in the mid-1800s, phrenology went trans-Atlantic, increasing its scientific sway in both America and Western Europe. By mid-century, American L.N. Fowler strengthened this authority not just in America but back in Europe as well through lectures and print where he furthered the construction of female domesticity and fragility through his similar, but elaborated, findings. Fowler cites the "phrenological rule, that length of fibre is an index of activity, while width denotes power" and proceeds to show how the female brain is narrower, making these neurological fibers longer and giving her a domestic disposition that bends toward maternity and kindness. The male, with his wider head, is predisposed to "have naturally more dignity, pride, resolution, independence, force and energy, together with greater mental power, growing out of the greater width of his head" (132). The female's "softness, mildness, delicacy and beauty, and all other womanly qualities" contrast with the male's "vigor, strength and robustness," ultimately making "[t]he situation of the female…dependent upon the male." The female's wider pelvis also contributed to her position of weakness in that it caused her step to be "less firm and decided" (128). Fowler further applies his method to other regions of the body. In discussing the Cephalic [sic] region, which contains the brain, he again emphasizes that men anatomically exhibit greater intelligence (130). He concludes that the primary differences between the sexes are: "The male has stronger intellect, will and propensities, together with greater forces and energy of character, while the female has the stronger moral sentiments and domestic feelings" (132). Apart from anatomically determining traits of each gender, Fowler proceeds to determine how these traits affect attraction, but only in respect to the male's attraction to the female:

> […] his best affections cling to her on account of her womanly qualities. Any thing masculine in her, excites his *dissatisfaction*, not to give the feeling a stronger name. And broad, square, shoulders *are* masculine… There is in them nothing of that delicacy, appeal for protection, and all subduing loveliness, which we instinctively attach to the word *feminine.* (114)

By attempting to define masculinity and femininity in phrenological terms, Fowler constructs them as scientifically-determined categories. This gives him the opportunity to inscribe Victorian sentimentalities regarding sex and domesticity noted above. "Feminine" is now not just prescribed by society, but is *anatomically* defined as delicate, lovely and desiring protection, as exhibited by physical traits like small shoulders. When Fowler writes that the absence of such slight features would create men's "dissatisfaction," he directly places women as objects that either satisfy or dissatisfy men. Through this objectification, he implies a physiological moral code, suggesting that women *ought* to satisfy men based on his standardized measures.

By analyzing the scientific language and claims of these 19th century phrenologists, it is evident that the authority of science allowed them to advance and inscribe conventional Victorian gender norms that in turn reinforced the seeming naturalness of the sphere of domesticity for women. Through the public lectures and texts of the authors above, the Victorian public consumed studies that today might seem quaint and outdated. But more importantly, these texts implied scientific bases for conceptions of femininity and masculinity that shaped the gendered thinking of Victorian society.

Gendering Eugenics

Not long after the rise in popularity in phrenology, a somewhat related proposal offered a more detailed connection between physiognomy and mental faculties. Ten years after the publication of Charles Darwin's *On the Origin of Species* in 1859, the author's cousin, Francis Galton, published his "enormously influential" work on the science of heredity and eugenic marriages (Burdett, 207). While eugenics has commonly been associated with Nazism from the mid-twentieth century, the "science of improving stock" (Galton, *Inquiries,* 25) gained popularity in the late nineteenth century, not only through Galton's writings but also through public events where Galton offered tests and measurements at his laboratory for a small fee (Gould, 108). In *Hereditary Genius*, the author described his methods for studying mental, moral and physical traits and how such studies might be used to determine which types of individuals should procreate and which should not. In this work, Galton attempted to determine the classification and value of individuals through composite portraiture and measurement of the body's faculties. While authors like Stephen Jay Gould have highlighted the undeniable classist and racist messages in Galton's work (108), one can see in the analysis below that eugenics also offered an argument that perpetuated nearly identical gender constructions that are shown in the phrenology texts above.

Galton began *Hereditary Genius* with a decisive distinction between genders. After explaining who he studied, how he classified individuals and his proper etiquette in omitting eminent men's names from the text, he proceeded to explain how it would have been inappropriate to "introduce the names of female relatives that stand in the same category" (3). The naming of individuals within his publication is seen as something of poor etiquette, but his particular care not to do so with women suggests some kind of female frailty in the public light. He also placed priority on male individuals, suggesting that the females whom he included in his study were only the relatives of the "eminent men." This positioned women in a dependent role, reinforcing their secondary status and suggesting that they were unworthy of study through their own merits.

As Galton proceeded to explain his scientific findings, he focused on quantifying the traits of the sexes through measurement and scientific observation. In one example, he found the numbers measuring judicial ability "go far to prove that the female influence is inferior to that of the male in conveying ability" (63). Emphasizing the authority and objectivity of numbers and scientific measurements, Galton made the argument for woman's inferiority, but through language and measurement that appear to supplant his gendered bias with concrete data.

Galton not only quantified female traits, but he characterized females largely as cogs in the machinery of society in that they reared the men of the world, seamlessly moving from his scientific measurements to a politically-charged agenda that valued productivity and achievement among men. In explaining the nature of a successful man of science, Galton claimed that "It therefore appears to be very important to success in science, that a man should have an able mother" (196). Since women are "blinder partisans [than men] and more servile followers of custom," the mother could obediently follow the necessary measures by which her son should be educated at home (196). This suggests that the son, not the daughter, plays a role of primary importance in the family and that it is he who promises scientific achievement and success in the public sphere. The role of mothers in this proposal further reinforces male primacy since they are illustrated as a necessarily subservient, but useful, device in male cultivation. This pushes both the absent daughter and the obedient mother into the domestic sphere, rendering them both almost invisible to the larger society.

Later in the text, Galton again discusses females as relatives of eminent men. He characterizes them as disadvantaged regarding marriage due to their scientifically-determined character:

> The only reasonable solution which I can suggest, besides that of inherent incapacity in the female line for transmitting the peculiar forms of ability we are now discussing, is, that the aunts, sisters, and daughters of eminent men do not marry, on the average, so frequently as other women… Again, one portion of them would certainly be of a dogmatic and self-asserting type, and therefore unattractive to men, and others would fail to attract, owing to their having shy, odd manners… which are disadvantageous to the matrimonial chances of young women. (328)

Galton's "solution" here offers a number of insights into how he uses his science to inscribe existing gender norms. He begins again by situating women in relation to the distinguished, primary "eminent men." Galton does not allow for the possibility of eminent women to exist independently of men, but instead places the female relatives of such men in a secondary, dependent role. He also assumes the necessity of marriage for women when he implies that the "matrimonial chances of young women" are something of great significance. To have such chances damaged by "disadvantageous" traits, he claims, is unfortunate. These traits of disadvantage illuminate another issue as well. For a woman to be "dogmatic," "self-asserting," "shy" or with "odd manners" inevitably makes her unattractive to men in Galton's view. In placing excessive value on women's attractiveness, Galton objectified women, suggesting their need to attract through passivity. These women with intelligence levels akin to their scientist relatives threaten to bring women and their offspring into the public sphere. Galton's desire for them to remain childless suggests his preference for a race of people where women remain non-intellectual and stay out of the public sphere of men. Women, instead, should remain home—and this sentiment of his underscores the importance he places on their primary childbearing role in society. Karl Pearson, fellow junior eugenicist of Galton, supported this rationale, but took it further with its implications pertaining to education and the government. In discussing the possible ramifications of women's emancipation, he worried about the possibility that studying might have "ill effects on women's child-bearing efficiency" (Pearson, *Ethics of Free Thought,* 371). He directly reinforced the separate spheres binary in saying that a woman's "home-duties are really her labour-contributing to the community" (434), further placing women as functional cogs valued only for their ability to reproduce and maintain the home.

As with phrenology, Galton's theory of eugenics offered a scientific proposal but with strong social and political purpose. Galton's scientific language treated females as secondary, inferior individuals who were designed to fulfill a functional value to the species by bearing children. In doing so, Galton inscribed Victorian notions of domesticity, maternity and femininity through the authoritative language, measurements, and evidence of science.

Domesticating Development

Unlike phrenology and eugenics, 19th century theories of biology and evolution did not offer such obvious social and political messages regarding race, class, and duty to society. Regardless, texts on these subjects still propagated similar gender norms as those shown in the texts above. Messages of domesticity and female inferiority can be traced through scientific writings on development, transmutation and evolution as is shown in the works of Robert Chambers and Charles Darwin.

While Charles Darwin's *On the Origin of Species* is the 19th century's best-known work on evolution, another highly popular work on evolution had already been published in 1844. Robert Chambers anonymously wrote and published *Vestiges of the Natural History of Creation and Other Evolutionary Writings*, exhaustively covering topics from the beginnings of the universe to the creation of life and evolution (Chambers' interest in human development can in some part be attributed to the work of George Combe in the 1830s and his work on phrenology, but Chambers resists making strong arguments about character and physiognomy in this text). The work "was a publishing triumph," selling extremely well with a public that was desirous of snippets of scientific information. It was heavily written on in "newspapers, popular weeklies, and heavy-weight quarterlies" and quickly "became the manifesto of the new science of progress" among the individuals external to the scientific community (Secord, "Introduction", xxvii). However, the text was highly criticized for its poor methods and unsound research by scientists who were suspicious of the anonymous work's methodology. Some even believed "feminine authorship offered the easiest explanation for the book's failings," underscoring the ubiquity of notions of female inferiority (Secord, "Introduction", xlii). In spite of its poor reputation among the scientific community, the book was nonetheless highly respected as a scientific text by the general public. Adrian Desmond describes its broad social and political appeal: "In an age rejecting aristocratic intervention and whim and looking for constitutional means of change, Chambers deliberately remade God in the image of a benign Legislator. He dressed up the issue to appeal to the middle classes—those who looked to the law, not noble patronage, for their advancement" (260). In addition to the text's political appeal in a time of class upheaval with the Chartist Movement, Secord argues that the text embodied a strong message of traditional Victorian domestic life that likewise appealed to the public as he highlights how "images of pregnancy, birth, childhood and the family were deeply embedded in the structure and language of the book" (Secord, "Domesticating Evolution", 276). As Secord argues, in order to make the threatening concept of evolution and transmutation palpable to the public, *Vestiges* showed that "the birth of a new species was no more or less to be feared than the birth of a child" (Secord "Domesticating Evolution" 276). The analysis below pushes Secord's claims further, showing that Chambers' work not only embodied messages of domesticity in order to make evolution more like life in the home, but that these messages also exemplified the reassertion of a sphere of domesticity.

After explaining his theories on the universe and creation at the beginning of the text, Chambers proceeds to analyze and classify various species. His study of the behavior of bees offers a clear example of Victorian gender norms. He first discusses the difference in male and female stages of arrestment: "that early one at which the female sex is complete… Four days more make it a perfect male" (216). Here, he suggests the female exists in a state less-developed than the male, supporting the idea of female inferiority through the lack of biological advancement. The "perfect male" is depicted as the final point of development (a claim that relies on the problematic assumption that evolution has a purpose), suggesting the female's arrested development renders her incomplete and inferior. The female bee is also described as a creature "destined to enjoy love, to burn with jealousy and anger, to be incited to vengeance, and to pass her time without labour," while the male is described as a responsible member of the community, defending public rights through his "laborious, industrious, patient, ingenious, skilful [sic]" character. Chambers even suggests that had the male bee's "ovaries been developed," it would have undesirably become an irrational creature, hurtful to the good of the community (215). While Chambers' claims are about a non-human species, he maintains: "Let it not be said that the phenomena concerned in the generation of bees may be very different from those concerned in the production of the higher animals. There is a unity throughout nature which makes the one case an instructive reflection of the other" (217). In his comparison of the behavior of bees to "higher animals," Chambers implies that he sees the characteristics of the male and female bee echoed in the human male and female. Here again is evidence of the separate spheres discussed by Davidoff and Hall—females are emotional, idle and domestically "without labour," whereas the male is characterized as part of the public sphere through his labor and industry.

More strikingly, perhaps, is how the father of evolutionary theory —Charles Darwin—inscribes gender traits and roles in a similar way. Darwin, best known for his work *On the Origin of Species*, was controversial in Victorian England not for the implications of his work on sex, but rather for what his work implied regarding humanity's place in the universe. However, in analyzing his discussion of sex in other species in *On the Origin*, we can also see that his work embodies a gendered bias similar to that of his contemporaries. In the course of his analysis of biology and reproduction, Darwin shortly turned to the differences between the sexes in a small section on sexual selection where he emphasizes the "struggle between the individuals of one sex, generally the males, for possession of the other sex" (Darwin, *Origin* 101). Lynda Birke argues that this small section and the entire mechanism of natural selection, "with its emphasis on 'survival of the fittest', the superiority of men over women" reinforced the norms of male dominance and female passivity through its scientific language (129). Darwin's theories were in no way immune to the gendered biases present in the above authors' works. Birke's observations highlight the implications of *Origin* regarding representations of gender in scientific texts, but the analysis below underscores further how domesticity and fragility are specifically represented in his later work.

In 1871 Darwin took his idea of sexual selection further in *The Descent of Man*. In this text, Darwin developed his theories of reproduction, mate selection and the biological traits of both sexes by focusing explicitly on the development of early hominids and the human species—something he had completely avoided in *Origin*—and he did so largely by highlighting physical and behavioral distinctions between the sexes. In his discussion of how each sex expends their energy differently, Darwin reduces both the male and female to particular tasks: "The female has to expend much organic matter in the formation of her ova, whereas the male expends much force in fierce contests with his rivals, in wandering about in search of the female, in exerting his voice, pouring out odoriferous secretions, etc." (239). While the male is portrayed as an ambulatory, aggressive, dominant sex with different natural tasks—akin to being out in the public sphere—the female is contrastingly depicted only through menstruation—a private affair—and is thus reduced to a sex solely concerned with reproduction. Darwin lists no other natural task or action for the female, suggesting that reproduction is her only role. He further reflects gendered norms in his language by regularly characterizing males as fighting for "possession of the females" (262), placing the female in a role of passivity, nearly as an object of prey. While it may be the case that different forms of courting occur among different species, it is Darwin's language of "possession" here that evokes the gendered norms repeatedly illustrated throughout these texts. An unscientific term at best, "possession" directly reflects the idea of women as property.

Aside from physical and behavioral traits, Darwin takes his science to evaluate intelligence as well:

> It is generally admitted that with woman the powers of intuition, of rapid perception, and perhaps of imitation, are more strongly marked than in man; but some, at least, of these faculties are characteristic of the lower races, and therefore of a past and lower state of civilization. The chief distinction in the intellectual powers of the two sexes is shown by man attaining to a higher eminence, in whatever he takes up, than woman can attain—whether requiring deep thought, reason, or imagination, or merely the use of the senses and hands. (Darwin, *Descent of Man,* 629)

Categorizing women as from a "lower state of civilization" again reflects the gendered biases present throughout all of these texts. Women are again placed in a position of inferiority in all aspects, both physical and mental as he characterizes all female thought, creativity, dexterity and sensory faculties as undeveloped.

Both Chambers' and Darwin's evolutionary arguments present theories of human history but woven within their proposals are word choices and claims that mirror the contrasting characteristics and spheres present in Victorian society. While neither author appears to overtly intend to make the social and political claims that we see in the texts of phrenology and eugenics, nonetheless these texts reflect Victorian sentimentalities regarding gender and reinforce them.

Conclusion

With the growing popularity of commodified scientific information, science itself gained a deified, unquestioned authority within Victorian non-scientist consciousness that has been inherited in much of Western thought up through today. Given science's undisputed status at the time, Victorian gender norms that prescribed behaviors and traits of females in the domestic sphere were reflected in scientific texts with little public alarm. The works of Gall, Fowler, Galton, Pearson, Chambers and Darwin, upon deconstruction, all reveal the presence of the ideological notions of female inferiority and domesticity. Interestingly while authors in phrenology and eugenics—two more overtly socially and politically-driven scientific ideas—used scientific prose to promote misogynist and racist agendas, Chambers' and Darwin's work, which appear much more descriptive than prescriptive, are imbued with similar messages regarding gender norms. In the case of these nineteenth century scientific texts, the authors both inscribe and contribute to existing Victorian conceptions of gender in terms of physicality and intellect; they are products of their time in this sense, but because of their influence and broad reach they perpetuated such gendered thinking in the public sphere as well. Presented in scientific terms, the realm of the domestic sphere and traits of femininity could be normalized and sanitized as part of the natural order of the sexes.

Bibliography

Birke, Lynda. 1986. *Women, Feminism and Biology*. Brighton: Wheatsheaf Books Ltd.

Blöde, Charles A. 2000. "Dr. F.J. Gall's System of the Functions of the Brain" in Nicholas J. Wade (ed.), *The Emergence of Neuroscience in the Nineteenth Century: Phrenology Examined*. London: Routledge/Thoemmes Press, pp. 1-88.

Burdett, Carolyn. 2002. "From *The New Werther* to numbers and arguments: Karl Pearson's eugenics" in Roger Luckhurst and Josephine Mcdonagh (eds.), *Transactions and Encounters: Science and Culture in the Nineteenth Century*. Manchester: Manchester University Press, pp. 204-231.

Chambers, Robert. 1994. *Vestiges of the Natural History of Creation and Other Evolutionary Writings*. Chicago: University of Chicago Press.

Clark, Anna. 2007. "Gender and Radicalism" in Kelly Boyd and Rohan McWilliam (eds.), *The Victorian Studies Reader*. New York: Routledge, pp. 191-206.

Cooter, Roger. 1984. *The Cultural Meaning of Popular Science: Phrenology and the Organization of Consent in Nineteenth-Century Britain*. Cambridge: Cambridge University Press.

Dale, Peter Allan. 1989. *In Pursuit of a Scientific Culture*. Madison: The University of Wisconsin Press.

Darwin, Charles. 1874. *The Descent of Man*. New York: American Publishers Corporation.

—. 1909. *On the Origin of Species by Means of Natural Selection, or the Preservation of Favoured Races in the Struggle for Life*. New York: P.F. Collier & Son.

Davidoff, Leonore and Catherine Hall. 2002. *Family Fortunes: Men and Women of the English Middle Class 1780-1850*. Revised Edition.

Milton Park: Routledge.
Davies, John D. 1955. *Phrenology Fad and Science*. New Haven: Yale University Press.
Desmond, Adrian. 2007. "Evolution Before Darwin" in Kelly Boyd and Rohan McWilliam (eds.), *The Victorian Studies Reader*. New York: Routledge, pp. 255-267.
Fowler, L.N. 1974. "The Principles of Phrenology and Physiology Applied to Man's Social Relations" in Charles Rosenberg and Carroll Smith-Rosenberg (eds), *Sex, Marriage and Society*. New York: Arno Press, pp. 5-135.
Gall, Franz Josef. 1835. *On the Functions of the Brain and of Each of Its Parts: With Observations on the Possibility of Determining the Instincts, Propensities, and Talents, or the Moral and Intellectual Dispositions of Men and Animals, by the Configuration of the Brain and Head*. Boston: Marsh, Capen & Lyon.
Galton, Francis. 1998. *Hereditary Genius*. Bristol: Thoemmes Press. [1869]
—. 1883. *Inquiries into Human Faculty and its Development*. London: Macmillan.
—. 1908. *Memories of my Life*. London: Methuen & Co.
Gay, Peter. 1969. *The Enlightenment: The Science of Freedom*. New York: W.W. Norton & Company, Inc., 1969.
Gould, Stephen Jay. 1981. *The Mismeasure of Man*. New York: W.W. Norton & Company, Inc.
Levere, Trevor H. 1980. "Coleridge and Human Science" in Marsha P. Hanen, Margaret J. Osler and Robert G. Weyant (eds.), *Science, Pseudo-Science and Society*. Waterloo: Wilfrid Laurier University Press, pp. 171-192.
Lewis, George Cornewall. 1849. *On the Influence of Authority in Matters of Opinion*. London: Savill and Edwards Printers.
Maxwell, Clerk. 1890. "Introductory Lecture on Experimental Physics." *Scientific Papers*. Ed. W.D. Nivens. New York: Dover Publications.
Pearson, Karl. 1892. *The Grammar of Science*. London: Walter Scott.
—. 1888. *The Ethics of Free Thought*. London: T.F. Unwin.
Postman, Neil. 1992. *Technopoly*. New York: Vintage Books.
Scott, Joan Wallach. 1999. *Gender and the Politics of History*. New York: Columbia University Press.
Secord, James A. 2007. "Domesticating Evolution" in Kelly Boyd and Rohan McWilliam (eds.), *The Victorian Studies Reader*. New York: Routledge, pp. 268-281.
Secord, James A. 1994. Introduction to Roert Chambers, *Vestiges of the*

Natural History of Creation and Other Evolutionary Writings. Chicago: University of Chicago Press.

Witemeyer, Hugh. 1979. *George Eliot and the Visual Arts*. New Haven: Yale University Press.

Yeo, Richard. 2001. *Science in the Public Sphere*. Burlington: Ashgate Publishing Company.

'THE EUGENIC VOTE': EUGENICS AND SUFFRAGE RHETORIC IN THE EDWARDIAN PERIOD

VÉRONIQUE MOLINARI

On 31st August 1911, as the issue of the postponement of the second Conciliation Bill by the government[1] was still being debated in Parliament and suffrage militancy was reaching its peak, an article appeared in the national organ of the main suffragist organisation (the National Union of Women Suffrage Societies). Its title, "The Eugenic Vote", brought together two apparently antagonistic terms, one essentially related to women's subordination to motherhood, the other to female emancipation. Its contents, however, was a forceful argument that, among the current controversies about "free feeding, vaccination, education, care of the feeble-minded, claims of the old", women's suffrage was essential to eugenic aims and that it was "significant", in this respect, "that this new outburst of eugenic energy [was] coincident with the new woman's movement" (Stevenson, 356).

Eugenics, "the study of agencies under social control which may improve or impair the racial qualities of future generations" in the words of its founder (Galton: 1909, 81), [2] had, by the turn of the century, become increasingly popular among British intellectual circles and, as already illustrated by a number of studies (Bland; Robb; Richardson: 2001, 2008; Jones: 1995), appealed to a large number of women. That some feminists among them should have adopted some of the very ideas that had been—and continued to be—used against female emancipation (including women's access to higher education and to political rights) and which they had tried to challenge remains, however, somewhat difficult to understand. Yet, this article, although noticeable because of its title, was far from being the only one to use eugenic arguments to support the cause of women's suffrage.

To understand this process, and to determine whether the adoption of eugenic arguments in suffrage rhetoric stemmed from genuine adherence to eugenic concerns or corresponded to a process of instrumentalisation, one needs to take a closer look at eugenic arguments and, more particularly, at the points of convergence with the feminist movement which made this reappropriation process possible.

Eugenics and the Women's Movement

Although eugenic theories appeared in the 1860s with the works of Francis Galton,[3] it was in the early twentieth century that, in the wake of the Great Depression of the 1870s and as a result of a combination of factors that seemed to point to the decline of Britain's economic and imperial pre-eminence, the movement truly gained a large following. The steady decline of the fertility rate that had taken place since 1851,[4] the 40% rejection rate among the urban military recruits that was reported during the second Boers War (Jordan, 168) and corroborating statistics that demonstrated the diminishing size, weight and health of the recruits were all taken up by both imperialists and eugenicists to raise the question of British military strength and racial efficiency. The looming threat of Germany, which was widely seen as a much more efficient nation, added urgency to the issue after 1902, as fears about a possible physical deterioration were coming to a climax. While no evidence for such a thing as a "degeneration of the race" or "physical deterioration" was found by the inter-departmental committee on physical deterioration that reported in 1904 (*The Times*, 29 August 1904), these fears contributed to a surge of interest in eugenics[5] and to its transformation from a relatively obscure, neo-Darwinist, statistically-based science into an organized propaganda movement[6] which influenced the social policies led by the Liberal government between 1906 and 1911 in the fields of free school meals, free medical care and old age pensions. On the grounds of a belief in hard heredity, as propounded by Galton, most eugenicists proposed to improve the biological traits of the British race through "positive" eugenics, which implied the promotion of procreation of those deemed "fit".[7]

Because they came to the fore in public debate among growing fears of the threat to national health and efficiency that the urban poor might represent, it is not surprising, then, that female demands for expanded educational, professional and political rights should have raised concerns about the impact of women's emancipation on the production and nurturing of the British race. The growing militancy of the suffrage movement after 1905 added fuel to the debate, all the more since the peak years of militancy (1907-1913) corresponded both to the years when the Liberal Governments of Bannerman and Asquith introduced the most important reforms of their welfare programme and to a period when eugenic publications multiplied (Saleeby: 1909, 1911; Murray Leslie: 1910-11; Scharlieb: 1912). Thus, the year the article entitled "The Eugenic Vote" was published, 1911, was the year the National Insurance Act was passed and when Caleb Saleeby and Murray Leslie respectively published *Woman and Womanhood* and "Woman's Progress in Relation to Eugenics".

The preoccupation of eugenicists with motherhood and the hereditary worth of families and classes certainly made them alert to the social characteristics of the women's movement and, in particular, to the involvement of middle-class women—presumably among the best candidates for motherhood. Central to the debate was the question whether women's demands for greater emancipation were compatible with the needs of the race. "There is no doubt that the new woman is a more interesting companion than her predecessor, and that she has made great progress in the arts and sciences, in trades and professions", Dr Murray Leslie wrote in 1911, "but the question of questions is—is she a better mother of the race?" (283). On this issue, not all shared the same view. For many eugenicists, who saw women as primarily created for the reproduction and nurture of the race and initially even positioned them essentially as "breeders" (in Galton's theories, women were to function as a medium for the transmission of "genius" from father to son), the vote, higher education or work were likely to divert middle-class women from their primary tasks of childbearing and childrearing and might even have an impact on their maternal capacities. Emancipated women would not only become unwomanly (and unattractive, therefore, to men of their "stock"), they would have less time, if married, to bear and raise their children, and would therefore leave to women of the lower grades of society the responsibility for the reproduction of the race.[8] Francis Galton thus voted against granting degrees to women in Cambridge in 1897 (Rowold, 52) and later became a committee-member of the Anti-Suffrage League.[9] He was later joined by several other eugenic opponents of women's suffrage. In "Woman's Progress in Relation to Eugenics", Dr Murray Leslie confirmed: "We wish for the fit, not the unfit, women of England to be the mothers of its future sons and daughters [...] it is a noteworthy fact that in Finland and Australia—two countries which enjoy women's suffrage- the birth-rate is almost the lowest in the civilized world". France and the United States were provided as examples of countries where women had access to education or took an active part in the business of the country and where the birth rate was low by comparison with Germany, where people, "in an overwhelming majority share the Kaiser's view that woman's place is in the home" (283, 291-292). Mary Scharlieb—praised by Murray Leslie as "one of the ablest of the modern of school of women doctors", even though a product of higher education herself, also warned against intellectual overexertion for women in her essay on *Womanhood and Race Generation* (7).

"Eugenic Feminism"

Others, however, took a different line—even though not radically opposed. That was the case of Havelock Ellis (1859-1939), Caleb Saleeby (1878-1940) and, to some extent, Karl Pearson (1857-1936). These preeminent members of the Eugenics Education Society believed that women were not simply vehicles but *agents* of reproduction who, to fulfil their mission, needed to be educated and economically independent (Soloway: 1990, chapt. 6), thus establishing a ground for what would become, in Saleeby's words, "eugenic feminism" (1912, 6). In that respect, they found an—unwilling—ally in British naturalist, anthropologist and biologist Alfred Russel Wallace (1823-1913)[10], who, although strongly opposed to eugenics himself, had, as early as 1890, defended the idea that women needed to be free from the artificial economic constraints that forced marital choices upon them for a "truly natural selection" to take place. In 1913, in *Social Environment and Moral Progress,* he reasserted his belief that economically independent, educated and politically active women would be in a stronger position to freely select their partner in marriage[11] and would naturally opt for the best men, leaving aside the weaker sort which they were frequently made to marry:

> Broadly speaking, I think we may trust the cultivated minds and pure instincts of the women of the future in the choice of partners. The idle and the selfish would be almost universally rejected. The coarse and sensual man, the diseased or the weak in intellect, those having a tendency to insanity or to hereditary disease, or who possess any congenital deformity, would rarely find partners, because the enlightened woman would know that she was committing an offence against society, against humanity at large, in choosing a husband who might be the means of transmitting disease of body or of mind to his offspring. Thus it will come about that the lower types of men, morally, and the physically diseased, will remain permanently unmarried, and will leave no descendants; and the advance of the race in every good quality will be ensured. ("Woman and Natural Selection")

Thus, at a time when the suffragette movement was at its peak, Dr Caleb William Saleeby, one of the most prolific eugenic writers in the pre-war period, wrote that he could see no good reason why women should not be granted the right to vote, going as far as saying he believed women's suffrage would be "eugenic":

> The present concentration of feminism in England upon the vote, sometimes involving the refusal of a good end—such as wise legislation—because it was not attained by the means they desire, and arousing all manner of enmity between the sexes, may be an unhappy necessity so long as men refuse to grant what they will assuredly grant before long. But now, and then, the vital matters are the nature of womanhood; the extent of our compliance with Nature's laws in the care of girlhood, whether or not women share in making the transitory laws of man; and the extent to which womanhood discharges its great functions of dedicating and preparing its best for the mothers, and choosing and preparing the best of men for the fathers, of the future. The vote, or any other thing, is good or bad in so far as it serves or hurts these great and everlasting needs. *I believe in the vote because I believe it will be eugenic* (my emphasis), will reform the conditions of marriage and divorce in the eugenic sense, and will serve the cause of what I have elsewhere called "preventive eugenics," which strives to protect healthy stocks from the "racial poisons," such as venereal disease, alcohol, and, in a relatively infinitesimal degree, lead. These are ends good and necessary in themselves, whether attained by a special dispensation from on high, or by decree of an earthly autocrat or a democracy of either sex or both. For these ends we must work, and for all the means whereby to attain them; but never for the means in despite of the ends. (1911, 12)

It is in fact very likely that the choice of "The eugenic vote" as a title for the article published in *The Common Cause* in 1911was motivated by Saleeby's essay, which had been published a couple of months before.

The fact is that while the eugenic movement needed the support of women as a whole to grow (its members did not hesitate to claim that "It is to the women of the country we must look in this great eugenic movement... Could anything be more philanthropic than to stamp out degeneracy? ", as Mrs Alec Tweedie wrote in the *Eugenics Review* in 1912)[12], the success of their aims depended to a large extent on the role played by those who were now taking part in the suffrage movement. Thus, for Saleeby, those whom it was most necessary to win over to their cause were clearly "the best women, those favoured by Nature in physique and intelligence, in character and their emotional nature, the women who are increasingly to be found enlisted in the ranks of Feminism, and fighting the great fight for the Women's Cause" and whom he invited to "furnish an ever-increasing proportion of our wives and mothers, to the great gain of themselves, and of men, and of the future" "instead of increasingly deserting the ranks of motherhood and leaving the blood of inferior women to constitute half of all future generations, shall on the contrary" (Saleeby: 1911, 8).

Montague Crackanthorpe, co-founder (with Sybil Grotto) and President of the Eugenics Education Society, shared Saleeby's belief that it was the new, emancipated woman both physically and mentally healthier as well as better educated and economically independent, that was to be encouraged to take on her responsibilities now that she had "warmed her hands at the fire of life" and "been permitted to eat of the tree of knowledge of good and evil". Woman's suffrage was not only inevitable but desirable, insofar as, once enfranchised, women could contribute through politics, to reforms that would strengthen marriage and the family (116-117).

By "winking at" the women's movement and opening the door to a possible compatibility between eugenic aims and feminist aims, some eugenicists thus made it easier for members of women's organisations to re-appropriate a discourse which had so far been dangerous (insofar as it implied a restriction of women's freedom of movement as well as of work and educational opportunities) but now could prove useful to serve their cause by bringing increasingly popular (pseudo-)scientific arguments into suffrage rhetoric and helping to legitimize their demands. The difficulties which the suffrage movement was facing at the time probably proved an encouragement to do so.

Re-Appropriating Eugenic Arguments

As widely illustrated in the works of Bland, Soloway and Richardson, eugenics certainly proved appealing to many women in the early twentieth century: almost half the membership of the Eugenics Education Society in 1914 were women, the London branch counted a majority of female members (Bland, 229) and the Eugenic Laboratory founded by Galton in 1904 also counted, in 1908, 5 women out of 14 members (Brown, 295-307; Rowold, 50). Many of these were active in the suffrage movement and, as Lucy Bland also notes, many served as "visiting lecturers" for the Eugenics Society to various organizations, including women's groups.

That eugenics should have appealed to women is not surprising. Through its fight against alcoholism and venereal disease, it seemed to offer the promise of a new morality. Social purists in particular, Bland explains, "saw in eugenics the potential of scientific validation and reinforcement of moral purity beliefs" (230). Probably as important was the fact that it placed woman—as a reproducer—in control and at the centre of the scheme as the topics of lecture of the Eugenics Education

Society illustrate ("Eugenics and Womanhood", "The New Woman and Race Progress", "Women and Eugenics", also positioned women centrally in the movement). Considering the limits it imposed to women's freedom of action in the interest of the race, that it should have appealed to feminists comes more as a surprise. Yet, by the turn of the century, the concept of women as "mothers of the race" had also become a recurring feature of the feminist discourse and of suffrage literature, even though the idea sometimes carried different meanings. Many feminists re-appropriated the notion that the welfare of the race depended on women and did not hesitate to resort themselves to (pseudo-)scientific arguments about reproduction to show why it was essential to the future of the race that women should gain access to the decision-making process. What had been used as an argument against the women's vote (taking part in politics would divert them from their domestic duties and harm their bodily maternal functions) was now turned to their advantage. Thus, members of both the NUWSS and the WSPU justified their demand for women's suffrage with the argument that women's participation to politics was necessary to ensure Parliament paid proper attention to the maintenance and survival of a healthy race[13] through the care of children, the sick and the elderly. Similarly, the argument of the fall in the birth rate was reversed so that it was no longer attributed to middle-class women's selfishness (a common accusation) but instead to women's reluctance to give birth to children who would be condemned to poverty and disease. Thus, concerns about the future of the race were used to ask for infant and child welfare programmes that involved well-baby clinics, pure milk centres, low-cost lunches and medical programmes in schools, educational centres for mothers and maternity homes. Finally, as for political power, access to higher education, which had been and remained decried by many eugenicists,[14] was now presented as a means to turn women into better mothers (which Mary Wollstonecraft had already argued in *A Vindication of the Rights of Women*) and, as a consequence, contribute to racial improvement.

The idea that women's suffrage was necessary to politics as it would bring a special expertise on motherhood and children matters was not a new one. Millicent Garrett Fawcett, who had been at the head of the NUWSS since the 1890s, had argued, ever since the beginning of her involvement in the campaign for women's suffrage, that women's knowledge and experience of motherhood, far from being an impediment to participation in politics, was on the contrary an excellent argument in its favour.[15] Eugenics, however, now provided suffragists with "scientific" arguments at a time when concerns about the future of "the race" (a word which became a recurring feature of suffragist literature) were running high. The basic argument of eugenic feminism—that if women were to bear the responsibility of "guardians of the race",[16] then society had to empower them, through the vote, to fulfil that role—could then prove both persuasive and pervasive, both among constitutional suffragists and suffragettes. Thus "The Eugenic Vote" published in the *Common Cause* linked women's suffrage with race preservation while some contributors to *The Vote* (the organ of the Women's Freedom League) presented race as a female instinct, predicting an acceleration of national decline and "race suicide" if women were not emancipated (Mrs Edward Francis, "Race Suicide" 21 January 1911, p. 152). Among militants suffragists, Christabel Pankhurst, also argued that women had "a service to render to the state as well as the home, to the race as well as the family" (*Votes for Women*, 21 January 1909, p. 208) and that, when they obtained the right to vote, they would contribute to improve "the condition of the children, the housing of the people… the care of the sick and aged, the preservation of the family" (1911, quoted in Garner, 50). More explicitly, Emmeline Pethick Lawrence, who had left the WSPU to join the Women's Freedom League the year before, claimed in 1913 that the suffrage campaign aimed at helping women "to rear a healthy race" (*Votes for women*, 5 August 1913, p. 9). In 1908, already, she had written, in the same magazine:

> without the vote and the equality symbolized by the vote, and the power given by the vote to women, to work out their own salvation, and to express what is their own conception of human life, these new powerful life-impulses can never be given, nor can the terrible physical and moral diseases from which the social body suffers to-day ever be cured. The life of the world cannot be renewed or purified until the Soul of the world's Womanhood is released. (January 1908, 49)

These ideas were also conveyed through fiction, in works such as those of Sarah Grand, social purity feminist, suffragist (she was President of the Writers' Suffrage League) and novelist,[17] who used her novels to advance the eugenic cause (Henson, 278) and depicted women as having the power to solve what she termed "the population difficulty" (quoted in Richardson, *Love and Eugenics*, p. 104). Echoing Karl Pearson's essay on "Woman and Labour" (1894, 561-77), she made her point of view even more explicit in an interview for *Humanitarian*:

> women are the proper people to decide on matters of population. Men have not managed to regulate either the population or the social question at all satisfactorily, and it would be well to give us a chance of trying what we can do. We could do much if we had the suffrage; the want of electoral

power cripples our efforts. (Tooley, 168)

In addition to "acting" in the interest of the race through legislation, enfranchised women, some suffragists argued, could simply "be" more favourable to the race in their choice of a partner. At that stage, the fact that ideas such as those of Alfred Russel Wallace were now being taken up and expounded by some prominent eugenicists made it easier to hold such a discourse. The emancipated new woman, as Eugenia Newmarch wrote in the *Englishwoman* in 1910, would be "instinctively attracted to the best sort of man and would choose him for her husband rather than the dullard or wastrel who had nothing but wealth or family connections to recommend him" (33-40). Not only would emancipated women be devoted to the improvement of the race, they would also be "eminently fitted" to reproduce it through eugenic marriages. The argument was shared by suffragists such as Alice Vickery Drysdale and Mabel Atkinson, who felt modern feminists were more attractive than their predecessors, more likely to be married, and less likely to marry the wrong man out of necessity. Feminism, Atkinson argued, was the ally of eugenics as it gave independent women the possibility to select their mates carefully and eugenically (Atkinson, 51-56).

When one looks at the other writings of some of the suffragists that have just been mentioned, it seems very likely, however, that the argument of a fit motherhood was used by some as a kind of Trojan horse to let eugenics enter the suffrage discourse while what was aimed at, in the end, was not fit motherhood but purer social standards for both men and women. Christabel Pankhurst's 1913 pamphlet, *The Great Scourge and how to end it,* in which male vices and the resulting venereal disease were held responsible for the degeneration of the race, is probably the best example of such an instrumentalisation of the eugenic discourse. Because men's immoral behaviour, by causing sterility and infant mortality, was responsible for the degeneration of the race, it should be reformed and men's moral standards raised to those of women. If men did not reform, Pankhurst wrote the following year, "the worst fears of the eugenicists will be fulfilled and the race bred entirely from inferior stock" (in *The Suffragette*, 1914). Giving women the right to vote would ensure that such a thing did not happen. Equal and higher moral standards were thus no longer a goal in themselves[18]—and the vote the instrument to reach it—but the means through which the race could be saved, which turned women's suffrage into an instrument at the service, no longer of the social purity movement but of the nation. Double standards of morality,[19] which had for so long been condemned by feminists, had thus been transformed from a feminist issue into a matter of national interest. This kind of discourse was probably made easier by the fact that the idea that venereal disease and prostitution (its main vector) as well as alcohol (which was another object of reform for many women campaigners because of its link with domestic violence) conducted to race suicide had already been defended by eugenicists such as Caleb Saleeby, who had listed them among what he termed "racial poison" (as it could be transmitted to the next generation) (Saleeby: January 1911, 30-52).

The context in which these articles were published was one of huge tension as far as the issue of women's suffrage was concerned. Between 1906 and 1910, the Liberal government had not given high priority to electoral reform (see Pugh, 1970) and, although private members' bills were introduced almost annually by supporters of the NUWSS, none had ever got beyond their second reading. Early in 1910, an attempt to obtain an agreed inter-party solution to the question had led to the creation of a Conciliation Committee[20] and the introduction of a new private member's bill[21] but, although it had passed its second reading by 299 votes to 189, the bill had not been allowed further time in the session. Faced with such opposition, members of the Women's Social and Political Union had stepped up their militancy: the first hunger strikes had begun in June 1909, and had led to the forcible feeding of imprisoned militants; in November 1910, a suffragette march to Parliament square had met with spectacular police brutality and degenerated into what came to be known as "Black Friday"; more recently, the 1911 census had been boycotted by suffragettes following calls of the Union (interestingly, the 1911 census was to include for the first time, an extensive enquiry into women's fertility) (White, 6). Considering the situation, adopting eugenic arguments to ask for the vote, and asserting that enfranchised women would contribute positively to the preservation of the race, could certainly prove a wise move as it followed the Liberal government's concern for national efficiency (see Searle) and could therefore contribute to curb the latter's reluctance to grant any measure of female franchise.

The Limits to the Strategy

If eugenics could appear compatible with an essentialist kind of feminism and appeal to social purity feminists, the danger which the adoption of its discourse represented did not escape those in the women's movement who held different views. Even though the ideology could be considered as empowering mothers, the fact remained that it implied some measure of control over women's bodies as a whole and, as such, could turn out to be a dangerous ally. As Richard Soloway concluded in

Feminism, Fertility and Eugenics in Victorian and Edwardian England, whatever eugenicists believed about "the woman question", they all supported the same view that "female emancipation [...] must not interfere with reproduction of numerous progeny" of the fit (1982, 141). As it was, even Karl Pearson, who publicly sympathized with women's demands for education, working opportunities and the vote, had declared in his inaugural speech to 1885:

> We have first to settle what is the physical capacity of woman, what would be the effect of her emancipation on her function of race-reproduction, before we can talk about her "rights", which are, after all, only a vague description of what may be the fittest position for her, the sphere of her maximum usefulness in the developed society of the future. The higher education of women may connote a general intellectual progress for the community or, on the other hand, a physical degradation of the race, owing to prolonged study having ill effects on woman's child-bearing efficiency. (Pearson, 1885)

For these reasons, while articles and essays defending eugenic ideas and associating feminist aims with eugenic aims multiplied between 1911 and 1913, so did those denouncing the movement and placing the freedom of the individual and women's rights to education above duties to the race. Dora Marsden, founder and editor of *The Free Woman*, and Helen Winter, a regular contributor to the magazine—both described by Swedish feminist Ellen Key as "extreme feminists"—were among those who chose to assert "personality in opposition to, instead of within, the race".[22] "As a freewoman [...] I care nothing for the continuance of the race nor the reproduction of any man," Helen Winster wrote; "my desire is to continue myself" (Helen Winter, "The individualism of motherhood, *The Freewoman*, 7 March 1912, p. 312, in Bland, 234).

Interestingly, this defence of the rights of the individual over the interests of the race, and the rejection of woman's reduction to her reproductive functions, were not only to be found among radical feminists but in some of the feminist and suffragist publications which, only a couple of months before, had not hesitated to put forward eugenic arguments to ask for the vote. In a 1913 article of *The Englishwoman*, Minnie Taylor thus argued:

> We are often warned that nothing but disaster can result if the good of the individual is placed before the good of the race. But what other than disaster can result if the good of the individual is sacrificed to the good of the race? The individual is more real than the species. If the welfare of the species is inimical to the welfare of the individual, then the species had far better die out. (in Rowold, 61)

Votes for Women also derided Saleeby's discourse on the "celebrated queen bee, who does nothing but breed", suggesting to produce, instead, "the lady spider—she eats her husband". "Would this not be a simple solution to the whole question?", the magazine provocatively asked (*Votes for Women*, 25 Oct. 1912). More vehemently, Beatrice Hastings denounced, in the pages of *The New Age*, "the cloying sentimentalization of motherhood", Rebecca West condemned, in 1913, in *The Clarion*, "the animal life that the Eugenics Society orders women to lead"[23] and Cicely Hamilton, in a contribution to H.G. Wells' *Socialism and the Great State*, summed up the issue in the following way:

> The entire question now at issue, not only between Woman and the State, but between Woman and Society in general, can be narrowed down to this: has she, like the other half of the race, a primary, individual, and responsible existence? or is she what may be called a secondary being […]? Is she, in short, a personality, or merely the reproductive faculty personified. (Hamilton, 226)

These examples tend to show that the women's movement was not united in its support of eugenics and that, even among those who chose to use eugenic arguments to ask for the vote, some were aware of the danger that those could represent and were not ready to accept the whole ideology unquestioningly. The fact is that the adoption of eugenics arguments by part of the women's movement undoubtedly added to the blurring of the lines that already existed (see Delap) between an essentialist kind of feminism and anti-feminism. That Charlotte Haldane, who described herself as a feminist and a suffragist—and is still often referred to as such, should have been able to accuse feminists of having "fomented 'sex antagonisms' by competing with men economically and refusing to conform [...] to the masculine ideals of sex-relationships" (Haldane, 168)[24] provides an excellent example of the confusion entailed.

Conclusion

The re-appropriation of the eugenic discourse by part of the suffrage movement in the pre-war period can certainly be understood as a strategic measure to secure the vote at a time when national efficiency was a priority for the Liberal government and opposition to women's suffrage ran high, leading increased militancy to be met in turn with increased violence. If many feminists chose to take up the eugenic discourse and put forward, as part of their campaign for suffrage, race preservation, racial purity and motherhood, this was partly because they had to, one of the most regular attacks made against the case for female emancipation being that it would lead to a less numerous and "weaker race." Although it makes no doubt that some suffragists were seduced by the rhetoric of the promise of social agency for women—whether what they had in mind was the interest of the race or equal (purer) social standards between men and women, it seems that eugenics was, more often than not, a means to an end. Whether this strategy worked is more than doubtful if one considers the failure of the new attempts, between 1912 and the outbreak of the war, to obtain votes for women.[25] As for the links between feminism and eugenics, even though they continued, in far less obvious way, into the inter-war period, they were still, in Clare Makepeace's words, more of a "flirt" than "a marriage of convenience" (Makepeace, 66).

Bibliography

Primary sources

Atkinson, Mabel. 1910. "The Feminist Movement and Eugenics", *The Sociological Review*, Volume a3, Issue 1, pages 51–56, January.

Crackanthorpe, Montague. 1907. *Population and Progress*, London, Chapman & Hall, ltd.

Edward Francis, Mrs. 1911. "Race Suicide", *The Vote*, January 21, p. 152-53.

Ellis, Havelock. 1912. *The Task of Social Hygiene.* Boston: Houghton Mifflin.

"Eugenics". *Eugenics Review*, vol. 9, January-June 1912, p. 857.

Fawcett, Milicent Garrett. 1890. "Home and Politics, An Address Delivered at Toynbee Hall". London: London Society for Women's Suffrage.

Galton, Francis (1865). *Hereditary Talent and Character.* Originally published in Macmillan's Magazine, 12, pp. 157-166. available at http://www.cimm. jcu.edu.au/ hist/stats/galton/macindex.htm

—. 1892. *Hereditary Genius: An Inquiry into Its Laws and Consequences.* 2nd ed. London: Macmillan.

—. 1909. *Essays in Eugenics.* London: Eugenics Education Society.

Haldane, Charlotte. 1927. *Motherhood and Its Enemies*. London: Chatto and Windus.

Hamilton, Cicely. 1912a. "Women in the Great State", in H.G. Wells et al., *Socialism and the Great State: Essays in Construction*. New York: Harper and Brother, pp. 219-248.

—. 1912b. In H.G. Wells et al., *Socialism and the Great State: Essays in*

Construction. New York: Harper and Brother, p. 226.
Hobson, J. A. 1910. *A Modern Outlook*. London: Herbert & Daniel.
Kinealy, Arabella. 2011. *Feminism and Sex Extinction*. London: Fisher & Unwin.
Murray Leslie, R. 1911. "Woman's Progress in relation to eugenics", *Eugenic Review*, January; 2(4): 282–298, pp. 283, 291-292.
Newmarch, Eugenia. 1911. "Women and the Race", *The Englishwoman*, April, pp. 33-40.
Pankhurst, Christabel. 1913. *The Great Scourge and how to end it, 1913.* London: Pankhurst.
—. 1914. "Concerning Damaged Goods", *The Suffragette*, 20 February.
Pearson, Karl. 1894. "Woman and Labour", *Fortnightly Review* (1894), pp. 561-77.
—. *The Life, Letters and Labours of Francis Galton*. Cambridge: Cambridge University press, 1914.
—. 2013. "The Woman's Question", 1885, in *The Ethic of Freethought and Other Addresses and Essays*. London: Forgotten Books. [1901]
Pethick-Lawrence, Emmeline. 1908. "What the vote means for those who are fighting the battle", *Votes for Women*, January, p. 49.
Report of the Inter-departmental Committee on Physical Deterioration, Vol. I, Report and appendix, London: His Majesty's Stationery Office, 1904, available on https://archive.org/details/b21358916 [consulted 12/08/20115].
Saleeby, Caleb William. 1910. "Racial Poison", *Eugenics Review* 2 (April 1910- January 1911), pp. 30-52.
—. 1909. *Parenthood and Race Culture: An Outline of Eugenics.* New York: Moffat, Yard.
—. 1911. *Woman and Womanhood: A Search for Principles.* New York and London: Mitchell Kennerley.
Scharlieb, Mary. 1912. *Womanhood and Race-Regeneration*. New York: Moffat, Yard.
Stevenson, D.T. 1911. "The Eugenic Vote", *The Common Cause*, 31 August 1911, p. 356.
The Times. "Physical Deterioration", 29/08/1904, p.13.
Tooley, Sarah A. 1896. "The Woman Question: An Interview with Madame Sarah Grand", *Humanitarian*, 8, p. 168.
Votes for Women. 21 January 1909, p. 208; 5 August 1913, p. 9.
Wallace, Alfred Russel. 1913. *Social Environment and Moral Progress*, New York: Cassell and Company.
Wintrhop Evans, Mrs. 1911. "The Mother as a Factor in Human Progress", *The Vote*, September 9, 1911, pp. 248-49.

"Woman and Natural Selection". Interview With Dr. Alfred Russel Wallace, *The Daily Chronicle,* 4 December 1893. To be consulted at: http://people.wku.edu/charles.smith/wallace/S736.htm.

Secondary Sources

Allen, Ann Taylor. 2000. "Feminism and Eugenics in Germany and Britain, 1900-1940: A Comparative Perspective", *German Studies Review*, Vol. 23, No. 3 (Oct., 2000), pp. 477-505.

Bland, Lucy. 1995. *Banishing the Beast: English Feminism and Sexual Morality, 1885-1914*. London: Penguin, 1995.

Brown, Ian. 1988. "Who were the Eugenicists? A study of the formation on an early twentieth century pressure group", *History of education*, 17, pp. 295-307.

Burdett, Carolyn. 1998. "The Hidden Romance of Sexual Science: Eugenics, the Nation and the Making of Modern Feminism" in Lucy Bland and Laura Doan (eds.) *Sexology in Culture: Labelling Bodies and Desires*. Chicago: The University of Chicago Press.

Davin, Anna. 1978. "Imperialism and Motherhood." *History Workshop Journal* 5, pp. 9-65.

Delamont Sara & Lorna Duffin. 2013. *The Nineteenth-century Woman: Her Cultural and Physical World.* London: Routledge.

Delap, Lucy. 2004. "Feminist and anti-feminist encounters in Edwardian Britain", *Historical Research*, vol. 77, no. 196 (May).

Durant, John R. 1979. "Scientific naturalism and social reform in the thought of Alfred Russel Wallace", *The British Journal for the History of Science* Vol. 12, No. 1 (March), pp. 31-58.

Farly Christopher and David Hodgson. 1972. *The life of Bertrand Russell*. Nottingham: Spokesman Books.

Fernihough Anne. 2013. *Freewomen and Supermen: Edwardian Radicals and Literary Modernism*. Oxford University Press.

Garner, Les. 1984. *Stepping Stones to Women's Liberty: Feminist Ideas in the Women's Suffrage Movement, 1900-1918*. Fairleigh Dickinson University Press.

Hall, Lesley. 2004. "Hauling Down the Doubles Standard: Feminism, Social Purity and Sexual Science in Late Nineteenth Century Britain", *Gender & History*, Vol. 16, N°1, April, pp. 36-56.

Henson, Louise, Geoffrey Cantor et al. 2004. *Culture and science in the nineteenth-century media*. Aldershot: Ashgate.

Jeffreys, Sheila. 1985. *The Spinster and Her Enemies - Feminism and Sexuality 1800-1930.* London: Pandora.

Jones, Greta. 1995. "Women and Eugenics in Britain: the case of Mary Scharlieb, Elizabeth Sloan Chesser and Stella Brown", *Annals of Science*, 51, pp. 481-502.

—. 1980. *Social Darwinism and English social thought: the interaction between biological and social theory*, Brighton: Harvester, 1980.

Jordan, Thomas Edward. 1992. *The Degeneracy Crisis and Victorian Youth*, State University of New York Press.

Love, Rosaleen. 1979. "Alice in Eugenics-Land': Feminism and Eugenics in the scientific careers of Alice Lee and Ethel Elderton", *Annals of Science*, Volume 36, Issue 2, 1979.

Makepeace, Clare. 2009. "To What Extent was the Relationship Between Feminists and the Eugenics Movement a 'Marriage of Convenience' in the Interwar Years?". *Journal of International Women's Studies*, 11(3), September, pp. 66-80.

Pugh, Martin. 1978. *Electoral Reform in Peace and War 1906-18*, London: Routledge.

Richardson, Angelique. 2008. *Love and Eugenics in the Late Nineteenth Century: Rational Reproduction and the New Woman*. Oxford: OUP.

—. 2008. "The life sciences", in David Bradshaw, *A Concise Companion to Modernism*, John Wiley & Sons, pp. 6-33, p. 20

—. 2001. "'People talk a lot of nonsense about Heredity': Mona Caird and anti-eugenic feminism", in *The New Woman in Fiction and in Fact: Fin de Siècle Feminisms*. London: Palgrave.

Robb, George. 1997. 'Race Motherhood: Moral Eugenics vs Progressive Eugenics, 1880-1920', in Claudia Nelson and Ann Holmes (eds), *Maternal Instincts: Images of Motherhood and Sexuality in Britain, 1875-1925.* London: Macmillan Press.

—. 1998. 'Eugenics, Spirituality, and Sex Differentiation in Edwardian England: the Case of Frances Swiney', *Journal of Women's History* 10.3: 97- 117.

—. 1996. 'The Way of All Flesh: Eugenics, Degeneration and the Gospel of Free Love'. *Journal of the History of Sexuality* 6: 589-603.

Rowold, Katharina. 2009. *The Educated Woman: Minds, Bodies, and Women's Higher Education in Britain, Germany, and Spain, 1865-1914.* New York and London: Routledge.

Searle, G. R. 1976. *Eugenics and Politics in Britain 1900-1914*. Leyden: Noordhoff International Publishing.

Soloway, Richard A. 1982a. *Birth control and the population question in England 1877-1930*. Chapel Hill.

—. 1982b. "Feminism, Fertility and Eugenics in Victorian and Edwardian England" in George Lachmann, *Political Symbolism in Modern*

*Europe,*Brunswick: Transaction, pp. 121-145.
—. 1990. *Demography and Degeneration, Eugenics and the Declining Birthrate in Twentieth Century Britain.* The University of North Carolina Press.
White, Ian. 2010. "No vote – no census: an account of some of the events of 1910–1911", in *Population Trends* nr 142, Winter. Office for National Statistics.

Notes

[1] A first Conciliation Bill to extend the parliamentary franchise to women householders had been introduced by a Labour MP, David Shackleton, in June the previous year and had passed its second reading by a large majority but Parliament had been dissolved before it could be enacted. A second Conciliation bill was introduced following the return of the Liberals to government and passed its second reading in May with an even larger majority. A few weeks later, on May 29, the Chancellor of the Exchequer announced that, owing to the conditions of business, the government could not allot the bill the amount of time it required but would do so in the next parliamentary session.

[2] The term eugenics was coined by Galton in 1883 from the Greek word "eugenes", ie "good in birth".

[3] Galton believed that "a man's natural abilities are derived by inheritance, under exactly the same limitations as are the form and physical features of the whole organic world." (Hereditary Genius, 1869); "if talented men were mated with talented women, of the same mental and physical characters as themselves, generation after generation, we might produce a highly-bred human race, with no more tendency to revert to meaner ancestral types than is shown by our long-established breeds of race-horses and fox-hounds" (Galton, 1865).

[4] From 304 births/1,000 married women aged fifteen to forty-five in 1851 to 234/1,000 in 1903 (Soloway 1990, 6-7).

[5] Among its recommendations, originating from Pr. D. J. Cunningham, was an anthropometric survey of the British population (Report of the Inter-departmental Committee on Physical Deterioration, 95-102).

[6] The Eugenics Education Society was created in the Winter 1907-8 under the impulse of Sybil Gotto to spread the new science and soon launched its own journal, *The Eugenics Review*. It could boast a fast-growing membership (had risen to 634 by 1914 and affiliated branches in Belfat, Birmingham, Liverpool, Oxford…) but, more important than the size of the Society's following, Searles notes, was "the intellectual caliber and the social prestige of its membership", which led to the Society being called to give evidence to the Home Office Inebriates Enquiry and before the Royal Commission on Divorce, and obtained from the government the setting up of a Royal Commission of Enquiry into Syphilis. Searle writes : "given the nature of the Society's work, and the hostility it aroused in some quarters, and the giggling embarrassment in others, it was

essential that the organization attract men of weight, gravity, and established reputation. By and large, it did so. Almost the entire biological establishment joined the E.E.S., and many of the most distinguished geneticists took an active part in its day to day work" (Searle, 11).

[7] "Negative" eugenics, which implied the active reduction of the fertility of the lower-classes through curtailment or prevention, was deemed at the time both politically and legislatively impossible. In 1913, the Mental Deficiency Act was to propose the mass segregation of the "feeble minded" from the rest of society but sterilisation programmes (even voluntary ones) were never legalized.

[8] William Ralphe Inge (later Dean of St Paul's), inaugural lecture to the Cambridge University Eugenics Society, 1911, quoted in Soloway 1990, 129.

[9] This decision proved an embarrassment for Karl Pearson, who wrote to Galton in 1908:"Among the fourteen workers in the Biometric and Eugenics Laboratories at present we have five women and their work is equal at the very least to that of the men. I have to treat them as in every way the equals of the men. They are women who in many cases have taken higher academic honours than the men and who are intellectually their peers. They were a little tried therefore when your name appeared on the Committee of the Anti-Suffrage Society" (Pearson was probably referring here to the Women's National Anti-Suffrage League which had been founded the same year). 15 December 1908, in Karl Pearson, *The Life, Letters and Labours of Francis Galton*, p. 359.

[10] "When men and women are alike free to follow their best impulses; when idleness and vicious or useless luxury on the one hand, oppressive labour and starvation on the other, are alike unknown; when all receive the best and most thorough education that the state of civilisation and knowledge at the time will admit; when the standard of public opinion is set by the wisest and the best, and that standard is systematically inculcated on the young; then we shall find that a system of selection will come spontaneously into action which will steadily tend to eliminate the lower and more degraded types of man, and thus continuously raise the average standard of the race". Human Selection. *Fortnightly Review* 48 (n.s.; 54, o.s.): 325-337 .Wallace, however, was himself an opponent of eugenics and opposed all efforts at social "improvements" by lending natural selection a "helping hand." "The world does not want the eugenicist to set it straight. Give the people good conditions, improve their environment, and all will tend towards the highest type. Eugenics is simply the meddlesome interference of an arrogant, scientific priestcraft" (quoted in John R. Durant, 31).

[11] "While she will be conceded full political and social rights on an equality with man, she will be placed in a position of responsibility and power which will render her his superior, since the future moral progress of the race will so largely depend upon her free choice in marriage. As time goes on, and she acquires more and more economic independence, that alone will give her an effective choice which she has never had before" (148).

[12] *Eugenics Review*, vol. 9, January-June 1912, p. 857, reproduced in Richardson: 2008, 6-33, p. 20. Scharlieb wrote that same year: "It is no exaggeration to say that on woman depends the welfare of the race" (7).

[13] "If women had parliamentary votes, they would try to alter the bad land laws which cause these bad housing conditions and result in such wicked waste of life… Give women votes to protect the children", *1907 election poster* (in Farly and Hodgson, 26).

[14] See Rowold. Dr Arabella Kenealy was to write in 1920 in the explicitly entitled *Feminism and Sex Extinction*: "The women intellectuals who have done great work have been women who inherited talents so far above the average, as spontaneously to have reached high mental levels, without need to have sacrificed those womanly traits which gave the noblest values to such work […]. The woman of average brain, however, attains the intellectual standards of the man of average brain only at cost of her health, of her emotions, or of her morale" (155).

[15] "The motherhood of women, either actual or potential, is one of those great facts of everyday life which we must never lose sight of. To women as mothers, is given the charge of the home and the care of children. Women are, therefore, by nature as well as by occupation and training, more accustomed than men to concentrate their minds on the home and domestic side of things. But this difference between men and women, instead of being a reason against their enfranchisement, seems to me the strongest possible reason in favour of it; we want the home and the domestic side of things to count for more in politics and in the administration of public affairs than they do at present" (Millicent Garrett Fawcett, "Home and Politics, An Address Delivered at Toynbee Hall", London: London Society for Women's Suffrage, 1890).

[16] Another member of the WFL and regular contributor to *The Vote*, writer and journalist Constance Hartley, also stressed woman's "racial duty" to breed, presenting her as "the Guardian of the Race-body and the Race-soul" (in Bland, 233).

[17] Author of *The Heavenly Twins* and eugenic short story "Eugenia: a modern maiden and a man amazed", in *Our Manifold Nature,* 1894)

[18] In 1896 already, discussing the female franchise in an interview published in *The Humanitarian* (a journal edited by Victoria Woodhull—staunch eugenicist and author, among others, of *The Rapid Multiplication of the Unfit* (1891), Sarah Grand declared: "our influence would be chiefly felt upon questions of morality, and would, I believe, tend to purify the political atmosphere" (Tooley, 164).

[19] What seemed to make it necessary to have access to the legislative process to fight the double standards of morality was the fact that these had, throughout the second half of the 19th century, been embodied in government policies and legislation, such as the 1857 Divorce Act (which made adultery by a wife grounds for a divorce while a husband's adultery was not sufficient grounds in itself and had to be combined with cruelty, desertion or some other offence for the wife to ask for a divorce), the Contagious Diseases Act of the 1860s or, again, in a 1870 Royal Commission report on the working of these Acts, which stated that "there is no comparison to be made between prostitutes and the men who consort with them. With the one the offence is committed as a matter of gain; with the other it is the irregular indulgence of a natural impulse" (Royal Commission upon the Administration and Operation of the Contagious Disease Acts: Volume 1, London:

HMSO, 1871, §60, quoted in Hall, 39).

[20] The Conciliation Committee was chaired by Lord Lytton as chairman and consisted of 25 Liberal MPs, 17 Conservative, six Irish Home Rulers and six Labour.

[21] Most of the women who qualified (about one million) were believed to be middle or upper class. These provisions were favourable to the Conservatives and did not have much Liberal and Labour support.

[22] Ellen Key, *The Woman Movement*, New York and London: G. P. Putnam's sons, 1912, quoted in Fernihough, 114.

[23] *The Clarion*, 4 April 1913, in Fernihough, p. 115.

[24] Sheila Jeffreys rightly refers to this essay as an "antifeminist classic" (*The Spinster and Her Enemies - Feminism and Sexuality 1800-1930,* London: Pandora, 1985, p. 174).

[25] In 1912, a third Conciliation Bill (Parliamentary Franchise (Women) Bill), failed to pass its second reading by 14 votes, largely because of the lukewarm attitude of the government and the government's own manhood suffrage measure—the Franchise and Registration Bill—introduced that same year was rapidly withdrawn, leading the suffragettes to resume and intensify militancy.

BARACK OBAMA'S POPULARIZATION OF SCIENCE: A POLITICAL STRATEGY?

GRÉGORY BENEDETTI

"Well I'm not a scientist […] But you know what? I know a lot of really good scientists at NASA, and at [the] National Oceanic and Atmospheric Administration (NOAA)[1], and at our major universities."[2] These are the words Barack Obama uttered in his State of the Union Address, in January 2015, as he was evoking the crucial issue of climate change. He intended to make it clear that heeding global warming, and addressing it, would be on his list of priorities for the remaining two years of his presidency. In saying so, he stated his desire to take into account the daunting realities that many scientists across the globe have been warning populations about for more than a decade. This statement also enabled Barack Obama to assert his position on science, in the aftermath of a painful electoral defeat in the 2014 midterm elections, when Democrats lost both houses of Congress against Republicans. Barack Obama seized the opportunity of the State of the Union Address, a ritualized speech which offers the president a vantage point on American political life, to draw an ideological contrast between the two major parties on global warming and climate, as well as in all science-related issues. It sheds light on a divide that has been one of the hot-button issues of U.S. politics for decades. As a consequence, not only does this anecdotal remark made by the president in a very codified speech tend to prove the permanent partisanship that seems to pull apart Democrats and Republicans when it comes to science, but it also highlights the importance of science in contemporary American political discourse.

As science is being dealt with in political discourse, this remark also illustrates the unconscious desire of Barack Obama to make a distinction between the notion of legitimacy and that of credibility. The American president tries to don his political discourse with a guise of humility as he seeks to tackle scientific matters in front of influential Representatives and Senators who have received their position of sovereignty[3] from the people. For if Barack Obama, as president, is legitimate in the sense that he has the political right to speak of a particular topic, or to decide upon it in a defined field, he seems well aware that his credibility remains more limited when evoking science, a subject which he does not necessarily have the capacity to speak of with accuracy and expertise (Charaudeau, 52).[4] Nevertheless, Obama makes sure he can use his status as chief executive to put forward a subject that has always been at the core of his political agenda, ever since his first presidential campaign in 2008.

This illusion of humility, and this will to separate legitimacy and credibility on scientific issues, can be regarded as the outcome of a long and progressive process that has resulted over the centuries in the evolution of the status of science. As said in the introduction to this volume, in the 17th and 18th centuries, science gradually emancipated in an intellectual sphere which had always been dominated by letters (philosophers, writers, historians, or philologists), up to the point that a supposedly autonomous "republic of science" or "empire of science" may have emerged.[5]

The evolution undergone in the following decades has thus forced anyone willing to speak about science to consider it as a specific type of discourse, which is determined by rules and codes that need to be taken into account, lest the orator might suffer from a deficit of legitimization that could be detrimental to his status. This approach bears much similarity with the one that is employed when dealing with the sphere of politics and political discourse. The latter also presents a set of characteristics that confers upon it a particular status and requires precise ways of handling and analyzing it. The political field is that which emerges when "questions of political action, its finality and its organization are dealt with simultaneously and in interaction" (Charaudeau, 12). As a consequence, political discourse is part of a pre-defined sphere in which codes rule the actions and the public speeches of politicians. Yet the political sphere is one that works in interaction with other fields and has to remain open. The fact that politicians represent citizens forces them to heed outsiders' requests and to prevent what would seem to be an enclosed functioning of the political field.[6] In this sense, the U.S. president epitomizes this aspect as he is the one in the limelight of the media, constantly making sure the link between public opinion and politics is not broken.[7]

We can observe from the outset that science and political discourse share a series of common features in the sense that they have both striven to acquire a more autonomous status. This quest has led toward their professionalization, at the same time as it has often resulted in their breaking up, which has often pushed these two spheres to cut themselves off from the rest of society, more positively for the former than the latter.

Indeed, if science, on the one hand, has long been considered as a field seeking constantly irrevocable truths, political discourse, on the other hand, is still very frequently perceived as one of illusion that arouses skepticism, if not rejection, from society. As said in the introduction to this volume, because of their constant necessity to convince and seduce potential electors, politicians may sometimes use political discourse in a deceitful way (Le Bart, 2003). While science partakes in a tradition of a rigorous logos, that is driven by reason and scientific truth, political discourse seems, on the contrary, to have been grabbed by ethos and pathos, the two remaining elements of Aristotle's rhetorical pattern: "it seems that political discourse [...] has progressively shifted from the space of logos to that of ethos and pathos, from the space of the content of arguments to that of their staging" (Charaudeau, 35).

If politicians, as illustrated in this book, can use science with a view of advancing a political agenda (disguising their intent with the cover of science's reputation of veracity and maximal symbolic capital in society), they can also contribute to shedding light on science by putting in the limelight scientific issues that are often isolated in the anonymous world of their experts. Politicians, in this case, act as ombudsmen who promote science by means of their position of authority and sovereignty, which is highly publicized by the media.

After almost eight years of a Democratic presidency, one can assert that it is in fact this kind of bilateral relation of cooperation that Barack Obama has decided to entertain with the world of science. In doing so, the 44th American president has clearly shown his desire to walk in the footsteps of his Democratic predecessors, some will say, with a view of showing the close links that the Democratic Party appears to have tied with science over the last decades, up to the point of being very often perceived by the media as the "pro-science" party. If the public tends to see scientists as neutral individuals as far as political opinions are concerned, 52% of scientists describe themselves as liberal.[8] These results debunk the myth of scientific neutrality on political issues, and give more fodder to the theory of the Democratic pro-science approach, even though others, especially on the conservative side, denounce this theory and try to prove it wrong. Barack Obama himself, as he came into office, strove to offer an unbiased vision of the relation between science and politics by claiming that his presidency would be marked by a non-ideological approach in any science-related matter, as he declared on April 27th 2009, in a speech delivered at the National Academy of Sciences Annual Meeting in Washington D.C.: "Under my administration, the days of science taking a back seat to ideology are over."[9] He had already declared a month and a half earlier: "we make scientific decisions based on facts, not ideology."[10] At that point, one may think that Barack Obama was referring particularly to the Bush Administration on climate change, an issue that came to symbolize the Republican president's anti-science policy. The rejection of the Kyoto Protocol, for instance, as early as 2001 when he came into office, became a marker of his position on climate change. It is interesting to point out that when he officially reasserted this environmental accord, he declared: "I've just met with senior members of my administration who are working to develop an effective and science-based approach to addressing the important issues of global climate change."[11] Similarly, George W. Bush's sexual education policy turned out to be one driven by ideological considerations, rather than scientific observations, if we consider that he underlined the crucial role of abstinence over other ways of avoiding sexually transmitted diseases, when he outlined his policy against HIV or early pregnancy.[12]

Consequently, when it comes to analyzing Barack Obama's political discourse on scientific issues, it seems pertinent to focus on the representation of science that the president offers to the public, through the lens of different official speeches he made in his first six years in office. The interest of the study lies in examining whether Barack Obama has adopted a post-ideological approach, in so far as he aimed at projecting a vision of science based on pragmatism, not on ideology, as he initially claimed when he came into office in 2009.[13] In this respect, it will be necessary to explore how Barack Obama navigated the tightrope of the political and scientific spheres in order to reconcile them after eight years during which George W. Bush had been constantly lambasted for his administration's supposedly anti-science positions.[14] It will thus be pertinent to observe whether Obama actually set the nation on a new path, and managed to reassure the scientific community through initiatives and policies that aimed at putting science back at the forefront of the political arena. Early in his presidency, for instance, the "Educate to Innovate" campaign can serve as a model to try to determine Obama's rhetorical and political pattern on science. Furthermore, the White House Science Fairs speeches will constitute cases in point to better formalize Barack Obama's vision of science through the use of an annual event that he has himself created, with a view of reuniting, not only science and politics, but also science and the younger generations. In doing so, we will demonstrate that Obama endeavored to considerably modify the image of science to make it more transparent and more attractive. To achieve these goals, the president regularly called upon U.S. national mythology to illustrate the fact that science has always been part of the American DNA, but he also intended to use his discourse on science as a way of putting forward economic and social issues which were part of his political agenda.

The Obama Presidency: Renewing the Relationship between Science and Politics

The relation between science and politics in the United States has always been one of ambiguity, mixing at the same time tensions and harmony, which reflects the complexity of making these two worlds work together. This observation seems particularly accurate if we look at the federal budget dedicated to science over the last fifty years. If spending for scientific and technological projects involving the defense and security (applied science) of the United States has been stable, if not increasing, over this period, funding by the Federal government for fundamental scientific research declined by 60% between 1967 and 2007 (Fischer, 2013). In this forty-year time span, there were only two Democratic presidents, Jimmy Carter and Bill Clinton, an observation that reinforces the analysis along which Republicans are definitely more reluctant to allocate funds to science issues.[15] In his eight years in office, between 2000 and 2008, George W. Bush did little to change the reputation of the G.O.P. in the public mind, and more particularly, in the scientific community. The Pew Research Center showed in 2009 that most scientists held a rather negative image of George W. Bush as his administration became famous for its tepid, if not completely non-existent, support to scientific programs of different kinds, even though conservative pundits, and some overtly Republican scientists, have tried to rehabilitate the image of the former president in the science community. In 2008, as the media were trying to analyze with some hindsight the policies of George W. Bush, many deplored the fact that science had been abandoned under Bush Jr. Kevin Trenberth, a climate scientist at the National Center for Atmospheric Research declared: "science has been seriously undermined by the censorship and alteration of testimony and news releases" (Britt, 2008).[16] The positions of George W. Bush raised so much concern in the scientific community that, at one point in his presidency, the Union of Concerned Scientists accused the president and his administration of discarding scientific reports and findings that went against his political plans, especially as fundamental research was concerned (Britt, 2008). To confirm this trend, we can also notice that during his presidency George W. Bush made the decision to tremendously reduce the scope of action of his science advisor, as he denied John Marburger the title of "assistant to the President" (Pielke Jr. and Klein, 2009).[17]

In this context, Barack Obama wasted no time to claim his support for the scientific community, first as a presidential candidate, then as president-elect. During the primary season, and later in the last months of the general election, Barack Obama dedicated a large amount of time to repeat that his presidency would mark a renewal for all science-related issues. For instance, on his official website, a large chapter entitled "Investing in America's future. Barack Obama and Joe Biden's plan for science and innovation", listed all the scientific and technological priorities of the Democratic candidate. The discourse emphasized the necessity to start anew as for the relation between politics and science: "Barack Obama and Joe Biden understand that critical national goals can only be met if we renew our commitment to science, technology and innovation".[18] The document also set the objective of: "restoring integrity to U.S. science policy." In fact, Barack Obama's attacks on George W. Bush were essentially focused on the ideological aspect. The discourse of the Democrat consisted in denouncing the ideologically-driven position of Bush on science: "we need to end the Bush administration's war on science where ideology trumps scientific inquiry and politics replaces expert opinion. As president, Barack Obama will lead a new era of scientific innovation in America."[19] Obama preferred to offer to the public, and to potential voters, the image of a president who would trust, rather than defy, the scientific community. It is worth underlining that the expression "war on science" was initially coined by a liberal commentator, Chris Mooney, whose writings specialize in scientific policies. In his 2005 book, *The Republican War on Science*, he stigmatized the position of Republicans to denounce the undermining, even the rejection of science, by the party which was then in control of the presidency as well as the House of Representatives and the Senate. Mooney astutely dug deep into the rhetoric of the president's "war on terror" to vigorously discredit Bush in the eyes of the scientific community. Mooney was then acclaimed by some as "one of the few journalists in the country who specialize in the now dangerous intersection of science and politics" (Rensberger, 2005).[20]

To further distance himself from George W. Bush, Barack Obama promised early on that he would "restore and enhance the status of the President's science and technology advisor." He did so immediately after he came into office by appointing a Harvard professor, John P. Holdren. Most observers underlined that the choice of President Obama had been made out of political concerns, since Professor Holdren has been known for his liberal activism.[21] He was thus picked to advance the agenda of Barack Obama and restore presidential credibility with the scientific community, proving once again, if necessary, that science and politics constantly overlap. It shows that Obama, as a politician, does not completely shy away from ideology, or at least from political leanings, when it comes to making decisions on how to proceed with his scientific policies. It cannot be hidden that the illusion of neutrality is combined with an obvious form of political reasoning, in spite of the President's statement: "I've charged John Holdren and the White House of Science and Technology Policy with leading a new effort to ensure that federal policies are based on the best and most unbiased scientific information. I want to be sure that facts are driving scientific decisions—and not the other way around."[22] Here again, President Obama makes sure that he projects the image of the Chief Executive whose decisions in science-related matters are driven by pragmatism. The renewing of the position of the President's science and technology advisor tends to prove that Barack Obama wanted to restore the authority of science in his political discourse. The nomination of Professor Holdren aimed at giving the president a support in the scientific sphere and at recreating the link between science and politics that George W. Bush had somehow broken.

Moreover, during his campaign, and as soon as he was elected, Barack Obama adopted a resolutely modern tone when it came to addressing scientific issues by concentrating on technology and innovation, two concepts that were clearly inseparable from his scientific rhetoric. In 2008, and in the first years of his presidency, this type of rhetoric was intended to turn the page of eight years of Republican leadership. Obama wanted to project the nation into a new era, toward a future that would necessarily be brighter, especially in the context of the economic and financial crisis that broke out in the midst of the campaign against John McCain. The scientific rhetoric served the purpose of supporting the political agenda of the candidate by undeniably showing signs of convergence. The rhetoric of science was thus very often associated with one of promise, offering thus a new future. Words such as "expand" and "enhance" were frequently employed by the candidate.[23] This became obvious in Barack Obama's Inaugural Address in January 2009, in which he simultaneously rejected the Bush years and projected himself toward a new era that would give science its due place in the political sphere. Obama clearly distanced himself from a discourse that belonged to the "imaginary world of tradition", to use Patrick Charaudeau's phrase, to rather resort to an "imaginary world of modernity" (163). For instance, Barack Obama declared in the first minutes of his Inaugural Address: "we will restore science to its rightful place, and wield technology's wonders to raise health care's quality and lower its costs. We will harness the sun and the winds and the soil to fuel our cars and run our factories. And we will transform our schools and colleges and universities to meet the demands of a new age."[24] This extract strikingly illustrates Patrick Charaudeau's theory on the imaginary world of modernity, as he speaks of a rhetoric where "the action of man and his capacity to transform the world through his thinking (conceptual creation), his hand (manufacturing creation) and the new tools that he produces (technological creation) are celebrated" in opposition to the discourse of tradition (167). This quote from Obama's speech resorts to the three types of creation described above, and seeks to reinforce the position of the new president, and his party, as allies of the scientific community. Moreover, this rhetoric also corresponds to the rather progressive approach chosen by the Democrats during the 2008 presidential elections, as Obama and other influential members of the party tried to tilt the ideology slightly to the center-left, to further isolate the conservative views of their predecessors and their opponents. In a context of crisis, this more progressive tone appealed to frustrated and disappointed voters who were looking for a new form of politics and a new path for the country.

Barack Obama again focused on scientific issues and policies later that same year. He unofficially launched his science agenda in his speech at the National Academy of Sciences in April 2009. He then aimed first of all at integrating his scientific policy in the long tradition of American presidents. This is partly why he reminded his audience that the National Academy of Sciences had been created by Abraham Lincoln, the man he has taken as a model and a source of inspiration throughout his political career. What can be noticed is that Obama used this speech to demonstrate, once more, that science was an important aspect of his political project, but he also used the opportunity to advance his agenda by mentioning issues that he had decided to integrate into the scientific sphere. Healthcare and global warming, for instance, were thus at the forefront of his program and had a central position in his address. Obama sought to combine his program with the preoccupations of the scientific community as a way of showing that science and politics were intrinsically intertwined. In front of a panel of scientists, who also happened to be American citizens, he declared: "we face more complex challenges than we have ever faced before: a medical system that holds the promise of unlocking new cures and treatments—attached to a health care system that holds the potential for bankruptcy to families and businesses; a system of energy that powers our economy, but simultaneously endangers our planet."[25] Obama further included science as part of a much wider political agenda as he linked it to the economy and the security of the nation: "Science is more essential for our prosperity, our security, our health, our environment, and our quality of life than it has ever been before."[26] In doing so, not only did Barack Obama try to regenerate science and restore it to a higher position, but he also politicized it.[27] Contrary to George W. Bush who, during his eight years in office was guided by his personal ideological positions and preconceptions toward science, which made him deny some evidence about global warming for instance, Barack Obama launched with this speech a long relation of collaboration, at least from a rhetorical standpoint, between politics and science. He aimed at replacing scientific results at the center of the political game with a view of using them to advance his broader more progressive political agenda. This political strategy became even more obvious as the U.S. president used the *American Recovery and Reinvestment Act* to actually provide basic research and science with more funds. Obama announced that the *Recovery Act* would also serve as a springboard to reenergize the scientific sector through the efforts, for instance, of the government, to help universities and private companies invest more into research. Again Obama wanted to put forward his commitment to science to draw a sharp contrast with his predecessor: "my administration is already providing the largest single boost to investment in basic research in American history."[28] Obama sought to demonstrate that the national economic interests and the interests of the scientific community could overlap and be treated in a synchronized, rather than completely distinct way.

These examples shed light on the desire of the American president, as he took office, to restore science's prominence in the list of his priorities as the new chief executive. In doing so, Barack Obama also had in mind to "presidentialize" and "politicize" science too. As a consequence, by the end of his first year in office, this approach enabled the Democratic president to elaborate a very clear road book of his science policies. In fact, Obama had never lost track of his desire to connect with special constituencies through and thanks to science. By restoring science's popularity among young Americans and minorities, mainly, and elaborating a rhetorical strategy, he aimed at using science to tackle social issues that would be part of his priorities toward the end of his presidency. In other words, science cannot be regarded as a neutral topic which is handled on its own by the president, but rather as a subject which has a real political potential. This very astute political and rhetorical strategy can be observed with the White House Science Fairs that Obama initiated as part of a much larger STEM (science, technology, engineering, and mathematics) initiative for the country.

The STEM Initiative and the White House Science Fairs: Obama's Politicization of Science

Barack Obama's vision for science had been made clear ever since the beginning of his 2008 presidential campaign. Once Obama became president, he was thus compelled to turn the promises he had formulated on the campaign trail into political actions. To accomplish that, he clearly expressed his position in 2009, at the National Academy of Sciences when he declared: "I want us to think about new and creative ways to engage young people in science and engineering, whether it's science festivals, robotics competitions, fairs that encourage young people to create and build and invent—to be makers of things, not just consumers of things".[29] This sentence served as a transition that signaled a new beginning in the American presidential conception of the role of science with the younger generations and it paved the way for Obama's future projects and initiatives such as the White House Science Fairs. Clearly, Barack Obama aimed at capturing the imagination of young Americans by offering a more positive, and less enclosed representation of science, as he invited the scientific community to better communicate: "And so today I want to challenge you to use your love and knowledge of science to spark the same sense of wonder and excitement in a new generation."[30]

To better materialize the renewal of the relation between science and politics, through political discourse, Barack Obama found an anchoring point through education, a field capable of both reuniting politics and science, and reconnecting science with Obama's core constituencies, youngsters and minorities (ethnic and social). This is why, in November 2009, Obama launched his "Educate to Innovate" campaign, in a speech that marked the beginning of the Federal government's project to reverse a dangerous trend that had been observed for decades in the U.S. It underlined the steep decline of American students in subjects like science and math, to the point that the United States had been surpassed by much smaller European countries, as well as other relatively more minor nations in terms of population and worldwide influence. Firstly, Barack Obama clearly announced his objective as he declared that this "Educate to Innovate" campaign should be directed toward "moving our country from the middle to the top of the pack in science and math education over the next decade."[31] Obama used a political discourse that was resolutely dynamic, and hopefully inspirational, as he invited younger Americans to project themselves toward a future where science would occupy a major role. As such, his speech was sprinkled with touches of patriotism that fully granted Obama the position of official American spokesman for science, which counterbalanced U.S. students' poor educational results: "reaffirming America's role as the world's engine of scientific discovery and technological innovation. And that leadership tomorrow depends on how we educate our students today."[32] This sentence came as an echo to one he had uttered in front of the National Academy of Sciences a few months before: "I believe it is not in our character, the American character, to follow. It's our character to lead. And it is time for us to lead once again."[33]

As mentioned earlier, Obama's desire to give science more importance was coupled with the objective of modernizing and popularizing scientific activities to attract new generations and new populations that had beforehand been somehow excluded from this field. First of all, Obama's STEM program consisted in economic measures that aimed at getting the private sector financially involved in this scientific renewal, especially through quasi-philanthropic activities that were similar to those observed during the American industrialization in the late 19th and early 20th centuries. Obama thus listed the companies that would participate in what was presented as a national mission to restore science. He did not hesitate to publicly express his gratitude toward corporations like Intel, Xerox or Time Warner Cable, in an effort to demonstrate that civil society could get involved, and that science was a priority in Americans' daily life.

Nevertheless, more interestingly in the context of studying the representation of science in political discourse, Barack Obama also announced in his "Educate to Innovate" speech, the organization of an annual White Science Fair, that he described as a gathering of "winners of national competitions in science and technology."[34] By setting up such a national event, and by wrapping it up with a guise of amusement, he nonetheless granted himself with an opportunity to offer his vision of science every year. He particularly wanted to signal his renewed representation of science to turn it into the epitome of both American educational excellence and fun, two apparently antinomic factors that would yet clearly strike a chord in young citizens' minds, as Obama declared: "Scientists and engineers ought to stand side by side with athletes and entertainers as role models, and here at the White House we're going to lead by example. We're going to show young people how cool science can be."[35] Obama thus discards old standards about science's supposedly professional isolation to assign it a new social mission. Interestingly with a black politician, when looking at this approach, one can draw a parallel with former African American political leaders who also pointed out the importance of scientific excellence and education to achieve social progress. Unconsciously, Obama inscribes himself in the footsteps of W.E.B. Du Bois and Booker T. Washington, for instance, who perceived science and education as potential driving forces toward the social elevation of blacks. In a similar fashion Obama made clear that his new science policy was also shaped with a view of advancing his social agenda, particularly with minorities: "And we're going to expand opportunities for all our young people—including women and minorities who too often have been underrepresented in scientific and technological fields."[36]

It is in 2010 that Barack Obama delivered his first speech at the White House Annual Science Fair, on the day of the first edition of what would become a tradition for the rest of his presidency. As Obama perfectly understood the media potential of this event, he shaped his political discourse in such a way as to turn these White House Science Fair addresses into ceremonial and ritualized speeches. These addresses were primarily aimed at inspiring younger generations of Americans to get into scientific and technological fields with a view of curbing the decline of the nation in this domain. Yet, in order to achieve this goal, Barack Obama was inclined to frame this project into a larger national goal, not only one exclusively reserved to U.S. youth. As a result, one can notice the rhetorical strategy of the president that consists in resorting to U.S mythology and national ideals. As such, one type of mythology in particular seems to fit very well into the scientific mould: the West and all the images that U.S. popular culture has produced and disseminated over time. Just like the mythology of the West, the imagery of science is one that is constantly oriented toward the future as a symbol of modernity. In the fashion of Frederick Jackson Turner's *The Significance of the Frontier in American History*, Barack Obama regularly insisted on the necessity for Americans to grab the potential of science because the future of the nation depends upon it. This was, for example, the case in 2012 when he said: "[...] America's best days are still to come […] when you compete in something like this, you're not just trying to win a prize today. You're getting America in shape to win the future […] You're making sure America will win the race to the future."[37] Just like Turner, for whom the West represented the future of the American civilization process, science is described by Obama as the engine of U.S. future domination and competitiveness on a worldwide scale. This link between the national and collective imagery of the West and Obama's representation of science had already been quite apparent in 2010. At that time, the president clearly tied science to the American DNA and intrinsic character, just like Turner had done when he wrote that the West had shaped U.S. singularity. Indeed Obama attached terms related to bravery, initiative, and boldness, to the theme of science, when Turner had resorted to analogous illustrations when he contributed to create the mythology of the Western pioneer.[38] Obama declared: "it [this country] was forged of stronger stuff, by bold men and women who dared to invent something new, or improve something; who took chances; who crafted and built and who tested our assumptions, and who believed that in America all things are possible."[39] In 2015, Barack Obama went even further with this analogy when he relied on the idea of pushing boundaries farther, with a speech that was imbued with the image of the conquest and discoveries, echoing what Turner had hinted at in the late 19th century: "We've always been obsessed with tinkering and discovering and inventing and pushing the very boundaries of what's possible. That's who we are. It's in our DNA […] And that's not just our past, that's also our future."[40] As we suggested earlier, this desire to connect science with a national myth, that of the West, partakes in this strategy of rehabilitating science and technologies in a period when these two spheres sometimes arouse skepticism, if not fear, among people.

Likewise, Barack Obama does not hesitate to refer to the national ideal by digging into the political mythology of the Founding Fathers and the origins of the country to demonstrate that science is quintessentially interconnected with the very idea of the United States.[41] In using this more popular and non elitist approach, the president distances himself from a vision of expertise, employed in other contexts, and integrates Americans of all backgrounds into his project, not only the scientific community or people susceptible to be attracted to scientific issues. Moreover, without openly naming it, he ties scientific initiative to the ideal of the American dream, a myth that speaks to every American, from every generation, creating a chronological and social common ground that unites all citizens: "We know that innovation has helped each generation pass down that basic American promise, which is no matter who you are, no matter where you come from, you can make it if you try."[42] The recourse to the American dream ideal and mythology can be perceived by the observer as a way of paving the way toward a more assertive discourse that emerged recently in Obama's rhetoric.

Indeed, as underlined earlier, it would be erroneous to consider the White House Science Fair speeches as anecdotal and neutral addresses aimed at entertaining young Americans and pleasing the scientific community. Under the cover of amusement, Barack Obama actually intends to make the most of these media appearances to emphasize his desire to promote science, on the one hand, and to use it indirectly as a stepping stone to further advance his political agenda. We can then point out the fact that Barack Obama's discourse in the White House Science Fair speeches has become much more political and assertive, over the years. The best example is undeniably the March 2015 address, the most "political" in tone and content, especially toward two key constituencies for the Democratic party: immigrants and ethnic and social minorities. Delivered in the context of a year-long situation of tension between the African American community and police officers (one may think of the tragic events in Ferguson), and in the midst of a very controversial immigration reform, the 2015 White House Science Fair speech may have struck a special chord in the political arena. This speech testifies to Obama's desire to act upon the issue of race which he had been expected to handle in a much more spectacular way based on the historical impact of his election as the first black president: "We want to make sure everybody is involved. We want to increase the diversity of STEM programs, as well. And that's been a theme of this science fair. We get the most out of all our nation's talent—and that means reaching out to boys and girls, men and women of all races and all backgrounds."[43] This instance demonstrates that science can be used as a springboard for dealing with social issues, even in a limited way, some will argue. Indeed, because of Obama's own feebleness, as well as because of a number of political restrictions that keep slowing down the process of fully integrating minorities, and especially blacks, in the United States, the Democratic president's record on race will necessarily remain symbolic and unfinished. His administration never really seized the historic opportunity of his election to actually craft comprehensive reforms on the racial issue. However, this record symbolizes the limits of a strategy employed by most black elected officials, a strategy which addresses race, nowadays, through a politicization of the question. Through the objective of the STEM project and the combination of science and education, the president shows once again that the fate of ethnic minorities in the United States is unquestionably linked to a more indirect political strategy, rather than one that would confront racial inequalities unswervingly. This is why, in 2015, Barack Obama used his speech at the White House Science Fair to address the issue of ethnic diversity and racial integration, during an event which, at first sight, seemed unrelated. Obama particularly underlined the lack of diversity in scientific fields, but his remarks could very well be interpreted as a more general analysis of the economic and social situation of minorities in the country. Even the magazine *The Root*, dedicated to African American culture, celebrated Barack Obama's effort to bring forth the issue of racial diversity in science through this 2015 event which saw a lot of projects presented by young students of color from all over the country.[44] This is also why Barack Obama linked the history of minorities to the one of immigrants, especially in the context of the very harshly debated immigration reform. He did not hesitate to praise the achievements of young immigrants: "Now, I should point out that, like several of the young people here, Anvita and Ruchi are first-generation Americans. Their parents came here, in part, so their kids could develop their talents and make a difference in the world […] And as President, I'm proud of you, because America is going to be stronger and smarter and healthier, and a much more interesting place because of you."[45] This general approach, which combines the destiny of racial minorities and immigrants, enabled Barack Obama to tackle the racial question by turning it into a national issue, not just one concerning African Americans. It partakes in this post-racial strategy,[46] which includes race into the larger political sphere.

As a consequence, we can see that the series of speeches the president delivered for the annual White House Science Fair presents a much stronger political, and even ideological potential than first meets the eye. Under the scientific cover, these addresses actually offer the president a vantage point to clarify the vision of science he defends, but they also represent a powerful opportunity to throw some hints regarding his political orientations in different areas, for instance reducing racial inequalities.

If George W. Bush was repeatedly criticized for his supposedly anti-science positions, Barack Obama seems, on the contrary, to have reversed this trend. From the first days of his 2008 presidential campaign on, and through his presidency, he has been willing to offer the image of a politician who was supportive of the scientific community. This approach tends to reinforce the public perception that in contemporary American politics, the Democratic party has become an ally of the scientific sphere, not only for the sake of science, but more importantly in order to advance a political agenda that, indeed, shares common objectives with a field that has long been perceived as exclusive and elitist. As a result, science is constantly used as a privileged lens to broach several issues which are more politically loaded. Science and politics appear as two spheres which have developed their own rules and their own codes, and yet which complement each other, depending on the status and the intentions of the orator. In the case of Barack Obama, it seems that science has very often been perceived as a stepping stone toward larger political designs such as economic or racial policies. This renewed sense of cooperation seems to back up the idea that the separation of science and politics appears to be an illusion as Adam Keiper suggests: "This is a longstanding inclination of modernity—to see science as something that ought not to be sullied by the soot of politics."[47] Here, with Obama, it is exactly the opposite as the President strives to connect the two fields and to show their perfect compatibility. With hindsight, Barack Obama has thus de-ideologized science, in the sense that his approach to science-related matters has not been guided by pre-conceived beliefs and ideas like George W. Bush but rather by pragmatism. He has chosen to politicize it as he fully grasped the wide range of political opportunities this generally acclaimed sector offers. Obama kept alive the desire to coordinate scientific issues with more political and social preoccupations, as the White House Science Fairs initiatives tend to demonstrate. As such, the first black president has reoriented, on purpose and with political objectives in mind, the approach of the chief executive toward science, and partially repaired a bond that may have been deteriorated by his predecessor.

Bibliography

Primary sources

Bush, George W. "President Discusses Global Climate Change." To be consulted at: http://georgewbush-whitehouse.archives.gov/news/releases/2001/06/20010611-2.html

Obama, Barack. "Inaugural Address." January 21, 2009. To be consulted at: https://www.whitehouse.gov/blog/2009/01/21/president-barack-obamas-inaugural-address

—. "Remarks by the President at the National Academy of Sciences Annual Meeting." April 27, 2009. To be consulted at: https://www.whitehouse.gov/the-press-office/remarks-president-national-academy-sciences-annual-meeting

—. "Remarks by the President on the "Educate to Innovate" Campaign." November 23, 2009. To be consulted at: https://www.whitehouse.gov/the-press-office/remarks-president-education-innovate-campaign

—. "Remarks by the President at the White House Science Fair." October 18, 2010. To be consulted at: https://www.whitehouse.gov/the-press-office/2010/10/18/remarks-president-white-house-science-fair

—. "Remarks by the President at the White House Science Fair." February 7, 2012. To be consulted at: https://www.whitehouse.gov/photos-and-video/video/2012/02/07/president-obama-speaks-white-house-science-fair#transcript

—. "Remarks by the President at the White House Science Fair." April 22,

2013. To be consulted at: https://www.whitehouse.gov/the-press-office/2013/04/22/remarks-president-2013-white-house-science-fair

—. "Remarks by the President at the White House Science Fair." May 27, 2014. To beconsulted at: https://www.whitehouse.gov/the-press-office/2014/05/27/remarks-president-white-house-science-fair

—. "State of the Union Address." January 20, 2015. To be onsulted at: https://www.whitehouse.gov/the-press-office/2015/01/20/remarks-president-state-union-address-january-20-2015.

—. "Remarks by the President at the White House Science Fair." March 23, 2015. To be consulted at https://www.whitehouse.gov/the-press-office/2015/03/23/remarks-president-white-house-science-fair

Secondary sources

Achouche, Mehdi. 2012. "Biopolitique et biopouvoir au XXIème siècle: les biotechnologies dans le débat politique américain contemporain." *Revue Représentations*, February.

Aristote. *Politique.* 1993. Paris : Gallimard.

—. *Rhétorique.* 1996. Paris : Grasset, Collection Le Livre de Poche [circa 329-323 BCE].

Benedetti, Gregory. 2014. "L'émergence d'une classe politique noire post-raciale, 2000-2012, le cas du nord-est des États-Unis: tensions, rupture, et continuité idéologique." PhD. Le Havre University.

Berezow, Alex & Campbell, Hank. 2012. "Are Democrats Really the 'Pro-Science' Party?" *Real Clear Politics*, September 10. To be consulted at: http://www.realclearpolitics.com/articles/2012/09/10/are_democrats_really_the_pro-science_party_115367.html

Boer, Deng, Richard Monastersky, Lauren Morello, Sara Reardon & Jeff Tollefson. 2015. "Obama budget seeks big boost for science." *Nature, News and Comments,* February 3. To be consulted at: http://www.scientificamerican.com/article/obama-s-budget-seeks-big-boost-for-science/

Bourdieu, Pierre. 2000. *Propos sur le champ politique.* Lyon: Presses Universitaires de Lyon.

Britt, Robert Roy. 2008. "Scientists Say Bush Stifles Science and Lets Global Leadership Slip." *livescience.com,* January 20th. To be consulted at: http://www.livescience.com/9574-scientists-bush-stifles-science-lets-global-leadership-slip.html

Charaudeau, Patrick. 2005. *Le discours politique. Les masques du pouvoir.* Paris: Vuibert.

Charaudeau, Patrick & Maingueneau, D. 2002. *Dictionnaire d'analyse du discours*. Paris: Seuil.

Cheneseau, Jean-Jacques. 1995. *Le langage du politique.* Poitiers: Méthodes et Stratégies.

Dawson, Michael. 2001. *Black Visions. The Roots of African-American Political Ideologies.* Chicago: The University of Chicago Press.

Fisher, Misha. 2013. "The Republican Party isn't Really the Anti-Science Party." *The Atlantic*, November 11th. To be consulted at: http://www.theatlantic.com/politics/archive/2013/11/the-republican-party-isnt-really-the-anti-science-party/281219/

Handelsman, Jo. 2015. "Obama Administration Aims to Erase Science Talent Deficit." *Scientific American*, March 23rd. To be consulted at: http://www.scientificamerican.com/article/obama-administration-aims-to-erase-science-talent-deficit/?print=true.

Haywood, Andrew. 2007. *Political Ideologies.* New York: Palgrave Macmillan (4th edition). [1992].

"Investing in America's Future. Barack Obama and Joe Biden's Plan for Science and Innovation." To be consulted at: http://pubs.acs.org/cen/news/pdf/FactSheetScience.pdf. 2008

Keiper, Adam. 2009. "Science and the Obama Administration." *The New Atlantis*, Winter. To be consulted at: http://www.thenewatlantis.com/publications/science-and-the-obama-administration

Kolbert, Elizabeth. 2015. "The G.O.P.'s War on Science Gets Worse." *The New Yorker*, May 6th 2015. To be consulted at: http://www.newyorker.com/news/daily-comment/gop-war-on-science-gets-worse.

Le Bart, Christian. 2003. "L'analyse du discours politique: de la théorie des champs à la sociologie de la grandeur." *Mots: les langages du politique*. La ville entre dire et faire, 72.

Malakoff, David. 2015. "Science Gets its Moment in Obama's 2015 State of the Union." *Science Mag,* January 26th. To be consulted at: http://news.sciencemag.org/funding/2015/01/science-gets-its-moment-obamas-2015-state-union

Passeron, Irène. 2008. "La République des sciences : réseau des correspondances, des académies et des livres scientifiques au 18ème siècle." *Revue du Dix-huitième siècle*, pp. 5-28. Available at: https://hal.archives-ouvertes.fr/hal-00361446v2

Pew Research Center. 2009. "Public Praises Science: Scientists Fault Public." July 9th. To be consulted at: http://www.people-press.org/2009/07/09/public-praises-science-scientists-fault-public-media/

Pielke, Roger Jr. & Roberta Klein. 2009. "The Rise and Fall of the Science

Advisor to the President of the United States." *Springer Science+Business Media B.V.* 2009. To be consulted at: http://sciencepolicy.colorado.edu/admin/publication_files/resource-2719-2009.05.pdf

"Power of the Progressive Economics: the Clinton Years." Center for American Progress. October 28, 2011. To be consulted at: https://www.americanprogress.org/issues/economy/report/2011/10/28/10405/power-of-progressive-economics-the-clinton-years/

Stolberg, Sheryl Gay. 2009. "Obama Puts His Own Spin on the Mix of Science With Politics." *The New York Times*. March 10th. To be consulted at: http://query.nytimes.com/gst/fullpage.html?res=9C07E0DB173EF933A25750C0A96F9C8B63.

United States House of Representatives Committee on Government Reform. Minority Staff Special Division Investigation. "Politics and Science in the Bush Administration." August 2003. To be consulted at: http://socrates.berkeley.edu/~scotch/science_policy/Bush_Admin.pdf

Notes

[1] The NOAA is the the National Oceanic and Atmospheric Administration, an agency that was created in 1970 under the supervision of the Department of Commerce. Its mission consists in providing information on the environment in general, and the oceans and the climate in particular, through regular weather forecasts and other collections of data.

[2] Barack Obama's 2015 State of the Union Address. https://www.whitehouse.gov/the-press-office/2015/01/20/remarks-president-state-union-address-january-20-2015 accessed on July 1st, 2015.

[3] The phrase "position of sovereignty" is used here following Patrick Charaudeau's definition in *Le Discours politique. Les masques du pouvoir.* where he writes: "Sovereignty is a matter of representation: when one represents, he speaks (or acts) in the name of the entity that exceeds us and that has delegated—temporarily—the power to speak in its name" [translated by the author] (Charaudeau, 53).

[4] In *Le Discours politique. Les masques du pouvoir*, Patrick Charaudeau draws a distinction between notions of legitimacy and credibility (52).

[5] René Sigrist noted that the use of the phrase spread from 1740 onward, even though it had already been observed in Jean de la Fontaine's fable, "Le renard anglais", in 1693 (338, 2008). The "autonomization" of science became a counterpoint to the Republic of Letters, and representatives of the Enlightenment, like Condorcet for instance, strove to create a transnational scientific community (Achouche, 106). Eventually, in 1830, the term "sciences" turned into "science", a linguistic change that testified to the complete emancipation of the field. It would give birth to the adjective "scientific", as Auguste Comte gave a full-fledged

definition to the concept in his book, *The Positive Philosophy*. René Sigrist summarizes Comte's definition as follows: "Science falls under a coherent system of objective knowledge elaborated along specific procedures (observation, experimentation, induction, mathematization …) and independently from any religious or philosophical system" (in Sigrist, 352).

[6] In *Propos sur le champ politique*, Pierre Bourdieu opposes the political and the mathematical fields: "those who are involved in the [political] game cannot play exclusively among themselves without referring to those in the name of whom they speak and in front of whom they are, occasionally, accountable" [translated by the author] (60).

[7] Marc Augé confirms this vision as he states: "the political intervention is thus a model of what could be called an enlarged set of rituals […] It abides by a certain number of formal constraints" (in Charaudeau, 41). Consequently, just like science, politics, and more precisely political discourse, appears to be an elite space that has been subject to an intense professionalization over centuries, as Christian Le Bart suggests: "democracy tends to reserve access to the public space to professionals of politics, who actually monopolize the right to speak about politics with the privilege of being heard, taken seriously into account, commented on, and sometimes even obeyed by" (Le Bart, 2003).

[8] Pew Research Center. Public Praises Science: Scientists Fault Public. July 9, 2009.
http://www.people-press.org/2009/07/09/section-4-scientists-politics-and-religion/ accessed on August 10, 2015.

[9] Remarks by the President at the National Academy of Sciences Annual Meeting.
https://www.whitehouse.gov/the-press-office/remarks-president-national-academy-sciences-annual-meeting accessed July 13, 2015.

[10] "Obama Puts His Own Spin on the Mix of Science with Politics."
http://www.nytimes.com/2009/03/10/us/politics/10obama.html?_r=0 accessed on July 4, 2015.

[11] "President discusses Global Climate Change."
http://georgewbushwhitehouse.archives.gov/news/releases/2001/06/20010611-2.html accessed on October 6, 2015.

[12] "Politics and Science in the Bush Administration."
http://socrates.berkeley.edu/~scotch/science_policy/Bush_Admin.pdf

[13] In this work we will use Andrew Haywood's definition of pragmatism and ideology: "Pramatism is a behavior shaped in accordance with practical circumstances and goals rather than ideological objectives" (3). "An ideology is a more or less coherent set of ideas that provides the basis for organized political action […] All ideologies advance a model of a desired future, a vision of the 'good society'" (11-12).

[14] Alex Berezow and Hank Campbell oppose the idea according which only the Democratic Party supports scientific enterprise in the United States. "Are Democrats Really the 'Pro-Science' Party?" *Real Clear Politics*, September 10, 2012.
http://www.realclearpolitics.com/articles/2012/09/10/are_democrats_really_the_pr

o-science_party_115367.html
[15] Jimmy Carter benefited from a Democratic majority in the House of Representatives during his four years in office, making it easier for him to have his budget approved by the legislative branch. Bill Clinton enjoyed the same situation in the first two years of his presidency. He later capitalized on his popularity and his political triangulation to reach a compromise with a Republican-dominated Congress. Under the Clinton Administration, the federal funding for science was increased, as proposed by the president. The National Science Foundation benefited from a 30% boost in budget despite six years of divided government during which the Democratic president negotiated the budget with a Republican-controlled House of Representatives. President Clinton also showed support for genetic research.
https://www.americanprogress.org/issues/economy/report/2011/10/28/10405/power-of-progressive-economics-the-clinton-years/ accessed September 11, 2015.
[16] Robert Roy Britt. "Scientists say Bush stifles science and lets global leadership slip."
http://www.livescience.com/9574-scientists-bush-stifles-science-lets-global-leadership-slip.html accessed July 1, 2015.
[17] Roger Pielke Jr. and Roberta Klein. "The Rise and Fall of the Science Advisor to the President of the United States." *Springer Science Business Media*. February 24, 2009.
http://sciencepolicy.colorado.edu/admin/publication_files/resource-2719-2009.05.pdf accessed August 8, 2015.
[18] http://pubs.acs.org/cen/news/pdf/FactSheetScience.pdf, accessed August 11, 2015.
[19] http://pubs.acs.org/cen/news/pdf/FactSheetScience.pdf, accessed August 11, 2015.
[20] Boyce Rensberger. "Science Abuse." *Scientific American*. September 24, 2005.
http://www.scientificamerican.com/article/science-abuse/, accessed June 25, 2015.
[21] Adam Keiper. "Science and the Obama Administration." *The New Atlantis*. Winter 2009. http://www.thenewatlantis.com/publications/science-and-the-obama-administration, accessed June 28, 2015.
[22] *Ibid.*
[23] http://pubs.acs.org/cen/news/pdf/FactSheetScience.pdf accessed August 11, 2015.
[24] Barack Obama's 2009 Inaugural Address. January 21, 2009.
https://www.whitehouse.gov/blog/2009/01/21/president-barack-obamas-inaugural-address, accessed June 26, 2015.
[25] Remarks by the President at the National Academy of Sciences annual meeting. April 2009.
https://www.whitehouse.gov/the-press-office/remarks-president-national-academy-sciences-annual-meeting, accessed June 20, 2015.
[26] *Ibid.*
[27] The term implies that Barack Obama integrated scientific issues into his broader political agenda in order to show that the two spheres should cooperate. It suggests

that Obama claimed not to be influenced by pre-constructed ideas or beliefs, but rather wanted to appear as a President whose actions were based on scientific results.
[28] *Ibid.*
[29] Remarks by the President at the National Academy of Sciences annual meeting. April 2009.
https://www.whitehouse.gov/the-press-office/remarks-president-national-academy-sciences-annual-meeting, accessed June 20, 2015.
[30] *Ibid.*
[31] Remarks by the President on the "Educate to Innovate" campaign. November 2009.
https://www.whitehouse.gov/the-press-office/remarks-president-education-innovate-campaign, accessed June 20, 2015
[32] Remarks by the President on the "Educate to Innovate" campaign. November 2009.
https://www.whitehouse.gov/the-press-office/remarks-president-education-innovate-campaign, accessed June 20, 2015.
[33] Remarks by the President at the National Academy of Sciences annual meeting. April 2009. https://www.whitehouse.gov/the-press-office/remarks-president-national-academy-sciences-annual-meeting, accessed June 20, 2015.
[34] Remarks by the President on the "Educate to Innovate" campaign. November 2009.
https://www.whitehouse.gov/the-press-office/remarks-president-education-innovate-campaign, accessed June 20, 2015.
[35] *Ibid.*
[36] Remarks by the President on the "Educate to Innovate" campaign. November 2009.
https://www.whitehouse.gov/the-press-office/remarks-president-education-innovate-campaign, accessed June 20, 2015.
[37] Remarks by the President at the White House Science Fair. February 2012.
https://www.whitehouse.gov/photos-and-video/video/2012/02/07/president-obama-speaks-white-house-science-fair#transcript, accessed June 13, 2015.
[38] In *The Significance of the Frontier in American History*, Turner wrote: "American social development has been continually beginning over again on the frontier. This perennial rebirth, this fluidity of American life, this expansion westward with its new opportunities, its continuous touch with the simplicity of primitive society, furnish the forces dominating American character."
[39] Remarks by the President at the White House Science Fair. October 2010.
https://www.whitehouse.gov/the-press-office/2010/10/18/remarks-president-white-house-science-fair, accessed June 13, 2015.
[40] Remarks by the President at the White House Science Fair. March 2015.
https://www.whitehouse.gov/the-press-office/2015/03/23/remarks-president-white-house-science-fair, accessed June 13, 2015.
[41] "You think about our Founding fathers—they were all out there doing experiments—and folks like Benjamin Franklin and Thomas Jefferson, they were

constantly curious about the world around them." Remarks by the President at the White House Science Fair. February 2012. https://www.whitehouse.gov/photos-and-video/video/2012/02/07/president-obama-speaks-white-house-science-fair#transcript, accessed June 13, 2015.

[42] Remarks by the President at the White House Science Fair. February 2012. https://www.whitehouse.gov/photos-and-video/video/2012/02/07/president-obama-speaks-white-house-science-fair#transcript, accessed June 13, 2015.

[43] Remarks by the President at the White House Science Fair. March 2015. https://www.whitehouse.gov/the-press-office/2015/03/23/remarks-president-white-house-science-fair, accessed June 13, 2015.

[44] Edwards, Breanna. "President Obama celebrates diversity with young scientists at the White House Science Fair." *The Root*. March 24, 2015. http://www.theroot.com/articles/politics/2015/03/white_house_science_fair_president_obama_celebrates_diversity_with_young.html, accessed August 10, 2015.

[45] Remarks by the President at the White House Science Fair. March 2015. https://www.whitehouse.gov/the-press-office/2015/03/23/remarks-president-white-house-science-fair, accessed June 13, 2015.

[46] Benedetti, Gregory. "L'émergence d'une classe politique noire post-raciale, 2000-2012, le cas du nord-est des États-Unis: tensions, rupture, et continuité idéologique." PhD. Le Havre University, 2014.

[47] Adam Keiper. "Science and the Obama Administration." *The New Atlantis*. Winter 2009. http://www.thenewatlantis.com/publications/science-and-the-obama-administration, accessed June 28, 2015.

WAS THERE AN ALTERNATIVE? ECONOMIC SCIENCE, POLITICAL DISCOURSE AND THE NEOLIBERAL REVOLUTION

PAUL AUERBACH, GILLES CHRISTOPH AND MARC LENORMAND

Exactly one year into her premiership, in May 1980, Margaret Thatcher famously declared that "there's no easy popularity in what we are proposing but it is fundamentally sound. I believe people accept there's no real alternative" (Thatcher, 1980).[1] The "soundness" of her approach was a leitmotiv of Thatcherite discourse and the word "sound" can be treated here as a functional substitute for "scientific".[2] Thatcher's oft-repeated conviction that hers was the only way forward served both to legitimize the neoliberal policies she pursued and to delegitimize rival policies, such as the "Alternative Economic Strategy" (AES from now on), originally proposed by left-wing academics as an economic and political counterweight to the austerity measures implemented by the Labour governments of 1974-1979. During the half-decade that followed Thatcher's 1980 pronouncement, neoliberalism became triumphant in most Western countries–either openly, as in the UK and in the US, or covertly as, in France or Germany–while the AES and a range of social democratic interventions in many countries were consigned to oblivion. Many observers left and right were then drawn to concur with Thatcher that there was no alternative to free-market capitalism and that her policies were indeed grounded in a sound model of how the economy works. This inevitably begs the question of the role played by the economic theories underlying neoliberalism and the AES in the respective success and failure of these political projects.

Some light on the matter can be shed by investigating the interactions of economic science and political discourse,[3] during the transformative decade of the 1970s. What makes this period particularly relevant for the analysis of the interplay between economic science and political discourse is that it witnessed the collapse of a previously dominant scientific-political paradigm, Keynesianism, and the rise of a challenger to the title, neoliberalism. From the 1930s to the 1970s, a broad range of policies and accompanying discourses over the need for government macro-management of the economy to maintain full employment–dubbed Keynesianism–had moved from the margins of economic science and politics to acquire the status of commonsense constructs,[4] which meant that activist scientific promotion and rationalizing was less needed to support them (Clarke, 1997). As Keynesian discourses and policies no longer seemed to offer adequate explanations for and, more importantly, adequate policy solutions to the crisis of British capitalism (Britton, 1991; Thompson, 2006), Keynesian common sense came to be challenged by neoliberalism, which followed the same intellectual and political trajectory from marginality to consensus that Keynesianism had gone through (Cockett, 1994; Dixon, 1998). Economic science can therefore be seen as playing a prominent role in pushing policy paradigm changes, before somewhat retreating to the background of policy debates as a number of scientific arguments solidify into new common-sense constructs.[5]

Neither neoliberalism nor the AES should be seen as the straightforward political expression of, respectively, neoclassical and post-Keynesian economics. These two schools of economic thought harbor tremendous diversity in their midst, spanning a whole range of economic analyses and policy proposals, from which no single discourse or program can be unequivocally derived. It is therefore best to approach neoliberalism and the AES as political endeavors which drew pragmatically on various strands of economic science to address the prolonged crisis encountered by the British economy over the course of the 1970s.[6] To reconstruct the articulation of these political projects rooted in economic theory, the focus will lie on the twin issues of inflation and the trade unions as conceptualized by proponents of neoliberalism and the AES and on the political and electoral impact of these opposing paradigms. In the process, a number of complementary hypotheses will be offered as to how science-based policy solutions come to be adopted, tested, and rejected or retained as policy tools and political agendas. First, however, it is necessary to take a closer look at the reasons why science is such an invaluable resource to politics.

The Relationships between Science, Politics and Discourse

This section offers a stripped-down elucidation of the uses to which scientific theories are put in political discourse.

As a general rule, the uses of science in politics tend to be *instrumental*–they are a means to an end. The key political end toward which scientific theories are mobilized is the accumulation of symbolic capital or, more specifically, the assertion and bolstering of *legitimacy*, that most elusive and precious resource for politicians. As Max Weber wrote, "every [domination] system attempts to establish and to cultivate the belief in its legitimacy" (Weber, 1978, 213). Cosmology offers an ideal-typical example of the ways in which the epistemic authority of science can be leveraged to legitimize political systems. Historian of science John Henry explains that during the Middle Ages and the modern period, the belief in correspondences–i.e. the idea that some parts of creation corresponded to others, revealing God's plan–meant that the order of society was expected to reflect the order of the cosmos. To elicit obedience, political arrangements had to be rooted in the natural order of God's creation, while unnatural arrangements were deemed unworkable. Aristotelian geocentrism was thus perceived as supporting the feudal system, in which authority was shared between the king and the nobility, since the sun and the planets, the respective symbols for the king and the nobility, occupied a common place among the heavenly bodies orbiting the earth. As feudalism gave way to monarchism from the Renaissance onward, it was similarly felt that the sun-centered universe pictured by the newly-influential Copernican heliocentrism lent support to the absolute power of the monarch over the gentry (Henry, 2002).

It can further be argued that scientific theories are mobilized to bolster two main species of political legitimacy–the legitimacy of politicians in their own eyes and their legitimacy in the eyes of others, or what we might term *conviction-driven* and *opportunity-driven* uses of science. Margaret Thatcher's embrace of Friedrich Hayek's ideas provides an example of conviction-driven uses of science:

> Another colleague had also prepared a paper arguing that the middle way was the pragmatic path for the Conservative party to take, avoiding the extremes of left and right. Before he had finished speaking to his paper, the new Party Leader [Margaret Thatcher] reached into her briefcase and took out a book. It was Friedrich Hayek's *The Constitution of Liberty*. Interrupting our pragmatist, she held the book up for all of us to see. "This", she said sternly, "is what we believe", and banged Hayek down on the table. (Ranelagh, 1991: ix)

Although Thatcher was a chemist by training, who had presumably been schooled into sifting through evidence and offering tentative conclusions, the forceful manner in which she proselytized her views from the moment she became the leader of the Conservative Party–as illustrated by the above incident—reveals that there was more of the self-styled "conviction politician" (Thatcher, 1981) in her than of the rational scientist, save for the mitigating factor that she felt her convictions to be "sound", i.e. grounded in what works.

"What works" is actually the catchphrase under which Tony Blair introduced evidence-based policy-making in the hope of circumventing sterile ideological controversies over governmental action. The following anecdote recounted by Richard Sennett, however, seems to indicate that his uses of science were more opportunity-driven than evidence-based:

> I wrote a book about inequality and respect. And one day, I was watching the TV news and I saw the Prime Minister say: "We're launching a new programme about respect based on the views of Professor Richard Sennett." And I thought: "Hum, what's that mean?" And what it really meant was a new programme of punishing young people who didn't show respect for the state. And when I ran into Blair a couple of days later I said: "You know, you've got it all wrong. Because my notion is that the state has to show more respect for the community, rather than to have more subservience." And he looked at me and he said: "But I pronounced your name correctly"—i.e. it's all media, it's all publicity. (Sennett, 2010)

In many cases, of course, the reasons why politicians adopt scientific theories are motivated by an equal mixture of conviction and opportunity—they perceive such a move to be in their broad political interest. Put differently, this means that most politicians do not dispassionately assess competing scientific theories on the basis of objective criteria (e.g. superior intrinsic merits) but select those that best suit their needs, however defined. It should also be noted that, more often than not, economic theories do not so much trigger conversions or inspire policies but rather supply *a posteriori* rationalizations for long-held beliefs—as was the case with Thatcher—or for a predefined course of action—as was the case with Blair. To take another example, John Maynard Keynes' advocacy of public works and the Roosevelt administration's public works program predated the scientific rationale offered for these measures in the 1936 *General Theory of Employment, Interest and Money*.

Finally, staking part of their legitimacy on the epistemic authority of science is a way for politicians to reduce the costs involved in their activities. In politicians' own eyes, believing in the scientific legitimacy of a specific measure or of a general blueprint for action lowers the cognitive costs of decision-making—politicians can have more confidence in the soundness of the policies they pursue. In the eyes of others, benefiting from a modicum of scientific legitimacy lessens for politicians the negotiation costs of ensuring factional allegiance within the party, of securing electoral support from voters, or of enforcing social compliance among the population–the expertise of decision-makers is enhanced among party-members, constituents and citizens. But claiming scientific legitimacy is not necessarily an efficient means for politicians to reduce their negotiation costs. Indeed, the fundamental contestability of the social world, which is rarely the object of consensual agreement, and the tentative nature of scientific knowledge, which is cause for almost unceasing controversies, mean that it is always possible for one paradigm to be opposed by another, as was the case with neoliberalism and the AES when it came to diagnosing and curing the economic crisis Britain underwent throughout the 1970s.

Economic Science and the Troubled Seventies

This section first examines the claims of neoliberal and AES economists to scientific objectivity before turning to the analyses and remedies they offered to the socioeconomic crisis that Britain underwent in the 1970s.

The Neoliberal Claims to Scientific Legitimacy

Economics has made strong claims to its scientific status ever since Adam Smith attempted to ground his "science of man" in the Newtonian method (Phillipson, 2010, 100-101). Just as Newton had unified terrestrial and celestial mechanics into the universal law of gravitation, from which all motion on earth and in the heavens could be deduced, Smith thought he had singled out a universal principle of sociability–sympathy–from which all human interactions proceeded.

The economics of Milton Friedman and Friedrich Hayek, the two leading neoliberal thinkers, similarly took their cues from the natural sciences.[7] In 1953, Friedman published a paper entitled "The Methodology of Positive Economics" which is considered to be "the most influential work on economic methodology of [the 20th] century" (Hausman, 2007, 145). Borrowing eclectically from the instrumentalist anti-realism of Ernst Mach, the logical positivism of the Vienna Circle and the falsificationism of Karl Popper, Friedman's methodology was fashioned after physics, especially quantum physics, in its emphasis on success at prediction, even at the expense of "unrealistic" assumptions. According to him, the scientific nature of economics derived from its ability to make verifiable–and, conversely, falsifiable–empirical predictions.

By contrast with Friedman, Hayek was weary of mathematical formalization, quantitative aggregation and statistical testing. He nevertheless claimed that the social sciences were empirical rather than aprioristic and adhered to a falsificationist epistemology. As he saw it, the models built by the social sciences were not accurate enough to predict when specific events allowed by a given theory would happen but could, however, make themselves falsifiable–hence scientific–by "forbidding" the occurrence of events precluded by the theory. In his final book, *The Fatal Conceit*, published in 1988, Hayek took inspiration not from physics–as economists are wont to do–but from biology to develop an evolutionary account of human civilization based on a theory of cultural selection akin to Darwinian natural selection. Hayek strenuously asserted that his evolutionary approach scientifically demonstrated the superiority of free-market capitalism over socialism.

The AES Claims to Scientific Legitimacy

There is a hard economic core to the AES. A key episode for the various analyses which converged in the AES was the failure of the economic agenda of the 1964-1970 Wilson governments, whose bold project of modernization through planning and intervention was shattered by successive runs on the pound, with desperate attempts at propping up the currency ending up in reluctant devaluation. There were three main strands to analyses aiming at explaining what had happened and what could prevent a repeat of this defeat of Labour plans at the hands of international markets.

The central figure is Stuart Holland, a left economist and sometime adviser to Labour Prime Minister Harold Wilson, who played a decisive role in shaping the AES. He developed an analysis which centered on the rise of multinational companies as a new mesoeconomic power which could frustrate attempts by governments at macro-managing the economy (Holland, 1975, 1978). These analyses converged with those stemming from other groups and traditions. In its annual *Review*, the neo-Keynesian Cambridge Economic Policy Group argued that the failure of British governments in macro-managing the economy and avoiding the dual ills of inflation and unemployment was due to excessive openness to the world economy and the resulting focus on and problems with the balance of payments (Hickson, 2005, 171). Finally, the Marxists who published the *New Left Review* and *Socialist Register* looked at the history of British capitalist development and debated the role of 19th-century imperial development and of the long-established political supremacy of City financial interests over northern industrial interests as the reason for the backwardness of Britain's industrial sector (Hobsbawm, 1969; Kilpatrick and Lawson, 1980; Nairn, 1979; Rowthorn, 1983). In all three cases, economic history was central, which is a major divergence from the quantitative dimension of Friedmanite monetarism, then the leading strand of neoliberal economics.

Scientific backing for the AES was perceived and projected by the trade unions as central to the legitimacy of their own economic claims. The trade unions themselves had been undergoing a rapid process of professionalization since the 1960s (Taylor, 1978). The Trades Union Congress (TUC) started publishing an annual *Economic Review* in 1968, which formed a central section of the annual report of the General Council. Individual trade union affiliates of the TUC increasingly recruited graduates from the economics, sociology and industrial relations departments of British universities into their new research departments, and hired teams of sympathetic academics to make the economic case for their wage demands and to contribute to alternative industrial plans for ailing companies. The presentation of AES policies at TUC congresses and Labour Party conferences was propped up by references to the works of left economists which supported the claims made by the left trade unionists.[8]

Neoliberal Analyses and Remedies

Although both Hayek and Friedman openly recognized that their wider advocacy of personal freedom was based on ethical rather than scientific grounds, as far as trade unions and inflation were concerned, they each claimed that their policy prescriptions were strictly derived from their economic analyses.

In his 1967 presidential address to the American Economic Association, Friedman made the startling prediction–inferred from his adaptive expectations twist on the quantity theory of money–that inflation and unemployment would soon start rising simultaneously, something which was thought impossible in light of the available empirical evidence, formalized in the famous 1958 Phillips curve. When the British economy became mired in stagflation from 1971 onward, Friedman's prediction was vindicated and his policy recommendations acquired a new legitimacy. Friedman demanded that the overarching goal of economic policy no longer be Keynesian full employment at the cost of rising prices but stabilization of the price level by means of a strict rule constraining the growth rate of the money supply, with unemployment left to settle at its "natural" level.

Although Hayek also subscribed to a version of the quantity theory, he insisted that Friedman's rule-bound monetarism was not enough because it ignored the inflationary pressures arising from the institutional rigidity of the British labor market. Indeed, the law granted British trade unions immunities protecting them from tort liability and allowed closed shop agreements enabling them to exercise monopoly powers over the labor market, which meant they could use the strike threat to frustrate government attempts at curbing inflation through money-supply constraints. With the British state being widely perceived throughout the 1970s as having lost all control over "ungovernable" unions, Hayek's analysis was felt to have been borne out by events–much like Friedman's. Hayek therefore publically campaigned for the stripping of the unions' legal immunities and the abolition of the closed shop system. Taken together, Hayek and Friedman's diverging approaches enabled them to deliver a two-pronged attack on inflation and the unions.

AES Analyses and Remedies

Building on their analyses of the problems associated with the openness of the British economy, the backwardness of British industry and the disproportionate power wielded by multinationals, the left economists offered a series of remedies: protectionism in the form of import controls; an industrial policy of investment to modernize British industry; increased state control over the decisions and administration of private companies, either in the soft form of planning agreements, or in the stronger form of the nationalization of top companies (Cambridge Political Economy Group, 1974; CSE London Working Group, 1980). These remedies were accompanied by calls for the continuation or revival of demand management to facilitate the modernization of industry.

When the AES was adopted by trade unionists, more emphasis was put on how this economic program could save and create jobs, as industrial jobs were destroyed by the successive recessions of the mid-1970s and late 1970s, and on how the encouragement of demand actually meant more money in the pockets of workers and their families.[9] As it was transformed from economic analyses and remedies into a full-fledged political agenda for the labor movement, the AES also gained a new dimension: the need for democratic control over the running of the economy was used as a central argument for the nationalization of top companies, while one of the trade union mottos of the 1970s, "industrial democracy", was incorporated into the AES agenda (Trades Union Congress, 1968, 531; 1970, 572-576).

What is striking here is that the two issues which were considered as central and interrelated by neoliberal advocates–inflation and the trade unions–were simply not regarded as key issues by AES economists. A number of them considered inflation simply not to be a problem, as long as wages could increase in line with inflation. Those who saw it as having a destabilizing impact attributed it to Britain's exposure to shocks in the world economy and to attempts by companies at recouping the consequences of wage increases (Glyn and Sutcliffe, 1972), and therefore suggested price control. As far as trade unions were concerned, they were not seen as obstacles to the working of the economy, but on the contrary as necessary counterweights to the pressures exerted by multinationals and finance capital, as precious advocates of industrial investment, and as essential tools in pushing for democratic control over the running of companies (Coates and Topham, 1972). Neoliberal economics and the AES were predicated on completely opposite premises, and therefore delivered opposite policy proposals.

The Political Fortunes and Misfortunes of Economic Theories

This section assesses the successful and less successful ways in which neoliberal and AES theories were articulated into a political discourse,[10] translated into political manifestos, used in internal party debates, and turned into soundbites that could resonate with the wider public and therefore be converted into electoral victories.

The Political Articulation of Neoliberalism

Neoliberalism was articulated into a political discourse to be consumed outside the narrow confines of academia by a number of people and institutions occupying center stage in the British public sphere–neoliberal think tanks, like the Institute of Economic Affairs and the Adam Smith Institute; politicians, such as Keith Joseph and Margaret Thatcher; and, most influentially at first, economic journalists (Parsons, 1990).

Economic journalists were the initial vector for the popularization of neoliberalism among the British public. Samuel Brittan from the *Financial Times* and Peter Jay from *The Times*, the most prominent economic agenda-setters in the British media, were very much impressed by the weight of empirical evidence in favor of Friedmanite monetarism and proclaimed its scientific superiority to Keynesianism. Over time, they were rallied by Norman Macrae from *The Economist* and Hamish McRae from *The Guardian*, who came to grudgingly acknowledge some validity to Friedman's theories. Collectively, the leading economic journalists contributed toward shifting elite opinion in the direction of monetarism. Peter Jay also had a hand in the crystallization of a new bipartisan neoliberal consensus. Indeed, he convinced his father-in-law James Callaghan, the Labour Prime Minister from 1976 to 1979, that monetarism was an ideologically neutral policy tool that could be used by governments whether left or right to macro-manage the economy. As a result, Callaghan declared that Keynesian demand management was wrong-headed during his 1976 address to the Labour Party annual conference (Callaghan, 1976), in a passage that was drafted by Jay himself, and monetarism became incorporated into the Treasury's econometric model in the late 1970s, preparing the ground for the Conservatives' monetarist policies of the 1980s.

In the process of popularizing neoliberal discourse, economic journalists, politicians and think tanks also participated in its structuration–giving it the specific shape that it came to assume in the British public sphere from the 1970s onward. Together, they formulated a syncretic variant of neoliberal economics that could bridge the gap between Friedmanite and Hayekian ideas. This successful synthesizing of the various strands of neoliberal analysis into a seemingly consistent political discourse is not an achievement that should be taken for granted–Arthur Seldon, the editorial director of the Institute of Economic Affairs (IEA), tellingly remarked that the debate over the completeness or incompleteness of Friedmanite monetarism "once caused a certain amount of division among the hierarchy of the IEA, almost to the point of splitting the atom" (Seldon, 1981, 101). However delicate the task, Brittan and Jay's articles in the quality press, Keith Joseph's speeches before the British public and Arthur Seldon's prefaces to IEA publications articulated a view of inflation and the unions which reconciled Hayekian and Friedmanite monetarism, Hayek's labor-market-based variant providing the institutional foundations that were missing from Friedman's original doctrine. The joint efforts of economic journalists, Conservative leaders and IEA directors went a long way toward laying the discursive bases for the neoliberal consensus that would dominate British politics from then on.

One last fundamental aspect of the political articulation of neoliberalism, beyond its popularization and structuration, was its actualization–turning ideas into a reality. To reshape British society in its image, neoliberalism had to be translated into a political program and, further, into economic policies, a task which the new leadership of the Conservative Party performed in spite of internal strife between the Heathian moderate "wets" and the Thatcherite radical "dries". The commitment to curbing both inflation and the unions was the central plank of Thatcher's 1979 electoral platform, alongside curbing government spending. As soon as it was elected, the Thatcher government proceeded apace with austerity policies aimed at bringing down inflation through tight monetary targets, high interest rates and government borrowing cuts. The structural reform of industrial relations was much more of a piecemeal process, stretching all the way from the 1980 Employment Act to the 1993 Trade Union Act, but union activities were eventually no less radically curbed than rising prices. In retrospect, Thatcher sympathizers consider low inflation and weak unions, combined with privatization, to be her enduring legacy to British politics (Blundell, 2007).

The Political Articulation of the AES

The AES, which originated in the writings of left economists from the left of the Labour Party and from the Communist Party of Great Britain, increasingly gained traction in the labour movement during the 1970s.

Its first major articulation was in *Labour's Programme for Britain 1973*, which formed the basis for the 1974 Labour election manifestos (Wickham-Jones, 1996, 69-73, 93-95). However, the complex structure and internal dynamics of the Labour Party contributed to neutralizing its impact on policy, once Labour was in government between 1974 and 1979. Indeed, in line with the principles of parliamentary democracy, the Parliamentary Labour Party has, since its creation, enjoyed a relative autonomy from the other components of the party–the constituency parties, the trade unions, the socialist societies–and even from internal bodies like the National Executive Committee as well as from the supposedly supreme body within the party, the annual conference (Minkin and Seyd, 1977).

This tradition of independence aided the Labour government's desire to delay or renege on the essential tenets of the AES–import controls, planning agreements, nationalizations, industrial investment and industrial democracy. The right of the party, which dominated the parliamentary party and therefore the Labour government, became convinced that wage increases were a key contributing factor to inflation, and that inflation was the major issue, reducing the profit margins of British companies and undermining the confidence of investors. The action of the Labour government from 1974 to 1979 therefore rested on the increasingly strict application of incomes policies, and on cuts in public spending (Ludlam, 1990).

In the second half of the 1970s, AES policies, which had formed the basis for the Labour Party blueprint for socialist advance and economic modernization, were transformed into an oppositional economic program, a platform for the left critics of the policies of the Labour government. Motions calling for AES-style economic measures were pushed at TUC congresses and Labour Party conferences by an alliance of left trade unionists and left Labour politicians. 1978 was an exception: as trade unionists expected a general election to be called in the autumn, a motion was passed at the TUC Congress, which simultaneously advocated support for the AES and declared support for the policies of the Labour government (Trades Union Congress, 1978, 543).

With Labour in opposition after its defeat at the 1979 general election, there was initially a return to the pattern of the early 1970s: an increasingly vocal left managed to impose the AES as party policy, and one may argue that this was initially seen more generally in the party as a way to put clear blue water with the toxic policies of the Conservatives in government, and formed the basis of the manifesto on which Labour fought the 1983 general election. However, after the Conservatives were returned to power, the right of the Labour Party used the election results to discredit the AES and advocate modernization and adaptation to the emerging Thatcherite settlement (Reid and Pelling, 2005, 153-163).

The Popular Reception of Neoliberalism

Any discussion of the popular reception[11] of neoliberalism can be nothing more than conjectural. In a prior publication, one of the authors of this chapter has suggested some sort of commonsensical understanding of neoliberal economic theories among the population.[12] One can actually find some trace evidence of this, as the following personal anecdote recounted by social scientist Mark Blyth illustrates:

> My father was a butcher. He never took a course in economics. Yet sitting in his car just before the 1987 British election he told me that he would not vote Labour. I inquired why he would not. Because, said my father, "once they get elected, Labour will spend all this money on creating jobs, which is fair enough, but it never works. It just means prices go up. They'll try it again and again and prices will go up and up. Then they will have spent all this money and there will be none left for the schools and the hospitals, so they will have to borrow. But because there is inflation, it will cost more money to borrow, which means there will be less money for everyone else. This means we will all have to pay more on loans and such things, so

> people will have less money to spend. The less people spend, the more the economy slows down, and so there are fewer people in work. If the Tories get in again, they'll cut taxes, people will spend more, and there will be more jobs". […] Both bemused and impressed, I asked, "So why does the money you spend that comes from a tax cut create jobs while the money spent by a Labour government creates inflation?" He sat for a minute, and then he said, "Because it does. Governments shouldn't do that kind of thing." (Blyth, 2002, vii)

Clearly, Blyth's father had come to believe in the scientific legitimacy of the neoliberal school of economics and was convinced that the policies it advocated were the way of the world.

One can also find trace evidence for the claim that Britons broadly adhered to the basic tenets of neoliberal economic analysis in the opinion polls registering the most important problem in the eyes of public opinion. It appears that inflation, rather than unemployment, was the overriding concern of the British public over the course of the 1970s (see Fig. 1), only to be overtaken by the unions during the months leading up to the 1979 general election (see Fig. 2).

When considering, in the words of a *Financial Times* journalist, that "all political history shows that the standing of a government and its ability to hold the confidence of the electorate at a General Election depend upon the success of its economic policy" (Watt 1968), there is then little doubt that the outcome of the 1979 vote critically turned on the perceived ability or inability of the Labour and Conservative parties to solve the inflation-union problem. With the "Labour isn't working" advertising campaign they put up during the winter of 1978-1979, the Conservatives were obviously trying to cash in on the widespread sentiment that Keynesianism was failing and that the government was mishandling the economic situation. This strategy seems to have paid off handsomely, since opinion polls do show that with every passing month toward election day the Conservative party was increasingly felt to be equipped with the best policies to turn the country around (see Fig. 3).

The Popular Reception of the AES

As was suggested before, AES-style policies were never implemented by British governments. Even if they became official Labour Party policies on several occasions, the Labour leaders were always careful to distance themselves from the radical measures advocated by their own party (Forester, 1979, 76-78). So there were no major national spokespeople for AES policies, except for Tony Benn (1975), the hero of the left in the 1980s, whose views were constantly attacked by the Conservative press as illustrating the extremism of the so-called "loony left".

Now, it was argued by the right of the Labour Party that since AES policies formed the basis of the 1983 election manifesto, the AES had been tested to destruction in that election, which returned a landslide Conservative majority. This would suggest that in terms of popular reception, the AES was plainly unpopular (Kavanagh, 2004, 324-325). But the results of the 1983 election can be interpreted in other ways: one could also point, for instance, at the huge anti-Thatcher vote, which was however split between the Labour Party and the Social Democratic Party formed by right-wing secessionists from the Labour Party.

In a sense, however, it is true that, with its attack on pretty much every single tenet of free-market capitalism and free trade, with academic support for it limited to a left fringe in British universities, and without any backing in the press (in staunch contrast with the neoliberal right), the AES faced an uphill struggle to reverse common sense constructs and create a new collectivist consensus around democratic control of the economy. In the end, it remained the agenda of relatively small groups on the left of the British movement.

Conclusion

The comparison of the relative fortunes and misfortunes of neoliberal and AES economic theories shows that, in the social realm, science has powerful electoral effects when it is *properly* mediated by political discourse.

For science-based views and policies to form the basis for a new scientific-political consensus, and become common sense constructs, a number of obstacles are to be overcome. Whereas political discourse often offers an overarching narrative about society and the economy, modern science tends to be about offering specific, localized answers to specific, localized questions. It means that coherent, all-encompassing science-based narratives can only be the result of unifying discursive operations. Such operations are made by brokers in academia, journalism and politics who see a political interest in constructing a science-based, authoritative narrative to justify or condemn existing policies. Indeed, times of objective and/or subjective crisis of the state and of the economy offer opportunities for the promotion of new scientific-political accounts which challenge existing policies based on solidified common sense constructs.

So unifying narratives, brokering and opportunities are a first series of prerequisites for science to shape political discourse significantly.

The success of science-based political discourses is also mediated by the existence of already established political organizations and institutional mechanisms for selecting policies: parties and elections. Becoming official party policy may be seen as a prerequisite to being tested in an election. However, the stories of the Labour Party—in which the AES was adopted as official policy but disowned by party leaders—and of the Conservative Party—in which the neoliberal Thatcher was initially elected alongside a team of moderate one-nation Tories—suggests that the elevation from policy proposal to party policy to government policy is not straightforward. In a sense, to be truly successful, science-based political discourses need to capture the imaginations of both party members and leaders and of the electorate. Scientific-political agendas are used as blows in internal party struggles, but the blows are all the more powerful if they can draw on popular support for policies. Interpretations and constructions of popular aspirations and party prospects are therefore central here, as are the agenda-setters in the media and the power brokers within the parties who play decisive roles in the promotion and rejection of scientific-political agendas.[13]

Figures

Figure 1: *What is the Most Urgent Problem Facing the Country (1970-1979)?*

Source: King and Wybrow (2001).

Figure 2: *What is the Most Urgent Problem Facing the Country (October 1978-March 1979)?*

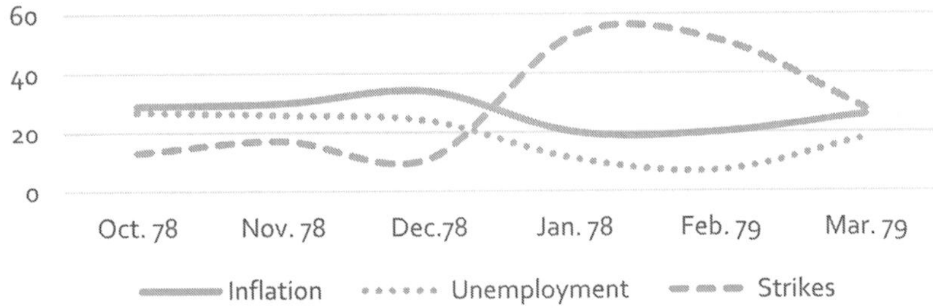

Source: King and Wybrow (2001).

Figure 3: *Which Party Can Best Handle that Problem (October 1978-April 1979)?*

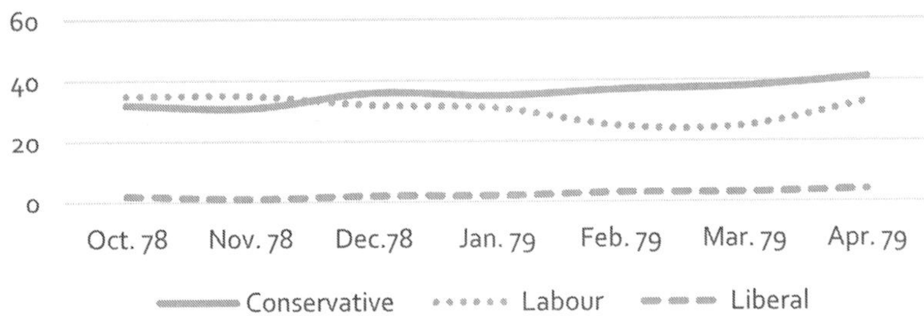

Source: King and Wybrow (2001).

Bibliography

Auerbach, Paul. 2007. "The Left Intellectual Opposition in Britain, 1945-2000: The Case of the Alternative Economic Strategy" in Francis King (ed.) *Is There no Alternative? Historical Problems of Socialist Economic Strategies*. Socialist History Society Occasional Papers No. 24. London: Socialist History Society.

Benn, Tony. 1975. *A New Course for Labour*. Nottingham: Spokesman.

Blundell, John. 2007. "Achieving Change: What We Can Learn from Margaret Thatcher". Heritage Lecture n° 1000. Washington, DC: Heritage Foundation, December.

Blyth, Mark. 2002. *Great Transformations: Economic Ideas and Institutional Change in the 20th Century*. Cambridge: Cambridge University Press.

Britton, Andrew. 1991. *Macroeconomic Policy in Britain 1974-1987.*

Cambridge: Cambridge University Press.

Callaghan, James. 1976. "Speech to the Labour Party Conference". Blackpool, 28 September.

Cambridge Political Economy Group. 1974. *Britain's Economic Crisis*. Nottingham: Spokesman.

Clarke, Peter. 1997. "The Keynesian Consensus and Its Enemies: The Argument over Macroeconomic Policy in Britain since the Second World War" in David Marquand and Anthony Seldon (eds), *The Ideas that Shaped Post-War Britain*. London: Fontana.

Coates, Ken and Tony Topham. 1972. *The New Unionism: The Case for Workers' Control*. London: Peter Owen.

Cockett, Richard. 1994. *Thinking the Unthinkable: Think-Tanks and the Economic Counter-Revolution, 1931-1983*. London: HarperCollins.

CSE London Working Group. 1980. *The Alternative Economic Strategy: A Response by the Labour Movement to the Economic Crisis*. London: CSE.

Dixon, Keith. 19998. *Les évangélistes du marché: Les intellectuels britanniques et le néo-libéralisme*. Paris: Raisons d'agir.

Dobry, Michel. 1986. *Sociologie des crises politiques*. Paris: Presses de la Fondation nationale des sciences politiques.

Forester, Tom. 1979. "Neutralising the Industrial Strategy" in Ken Coates (ed.), *What Went Wrong*. Nottingham: Spokesman.

Glyn, Andrew and Bob Sutcliffe. 1972. *British Capitalism, Workers and the Profits Squeeze*. Harmondsworth: Penguin.

Hall, Peter. 1990. "Policy Paradigms, Experts, and the State: The Case of Macroeconomic Policy-Making in Britain" in Stephen Brooks and Alain-G. Gagnon (eds.) *Social Scientists, Policy, and the State*. New York: Praeger.

—. 1992. "The Movement from Keynesianism to Monetarism: Institutional Analysis and British Economic Policy in the 1970s". *Structuring Politics: Historical Institutionalism in Comparative Analysis* in eds. Sven Steinmo, Kathleen Thelen and Frank Longstreth. *Cambridge Studies in Comparative Politics*. Cambridge: Cambridge University Press.

Hall, Stuart. 1973. "A World at One with Itself" in Stanley Cohen and Jock Young (eds.) *The Manufacture of News: Social Problems, Deviance and the Mass Media*. London: Constable.

Hausman, Daniel, ed. 2007. *The Philosophy of Economics: An Anthology* (third ed.). New York: Cambridge University Press, 2007. [1984]

Henry, John. 2002. *The Scientific Revolution and the Making of Modern Science*. London: Palgrave Macmillan. [1997]

Hickson, Kevin. 2005. *The IMF Crisis of 1976 and British Politics*. London: Tauris Academic Studies.

Hobsbawm, Eric. 1969. *Industry and Empire*. London: Weidenfeld & Nicolson.

Holland, Stuart. 1975. *The Socialist Challenge*. London: Quartet.

—. 1978. *Beyond Capitalist Planning*. Oxford: Blackwell.

Kavanagh, Dennis. 2004. "Conclusion: Reading and Misreading Old Labour" in Anthony Seldon and Kevin Hickson (eds.) *New Labour, Old Labour: The Wilson and Callaghan Governments, 1974-1979*. London: Routledge.

Kilpatrick, Andrew and Tony Lawson. 1980. "On the Nature of the Industrial Decline in the UK". *Cambridge Journal of Economics* 4:1.

King, Anthony and Robert Wybrow. 2001. *British Political Opinion, 1937-2000: The Gallup Polls*. London: Politico's.

Kuhn, Thomas. 2012. *The Structure of Scientific Revolutions*. Chicago: University of Chicago Press (4th ed.). [1962]

Ludlam, Steve. 1990. "Labourism and the Disintegration of the Postwar Consensus: Disunited Trade Union Economic Policy Responses to Public Expenditure Cuts, 1974-1979". PhD thesis, University of Sheffield.

McAdam, Doug, Sidney Tarrow and Charles Tilly. 2001. *Dynamics of Contention*. Cambridge: Cambridge University Press.

Merton, Robert. 1968. *Social Theory and Social Structure.* New York: The Free Press, 1968 (3rd ed.). [1949]

Minkin, Lewis and Patrick Seyd in William Paterson and Alastair Thomas (eds.) 1977. "The British Labour Party". *Social Democratic Parties in Europe*. London: Croom Helm.

Nairn, Tom. 1979. "The Future of Britain's Crisis". *New Left Review* 13/14.

Parsons, Wayne. 1990. *The Power of the Financial Press: Journalism and Economic Opinion in Britain and America*. New Brunswick, NJ: Rutgers University Press. [1989][

Phillipson, Nicholas. 2010. *Adam Smith: An Enlightened Life*. London: Penguin.

Ranelagh, John.1991. *Thatcher's People: An Insider's Account of the Politics, the Power, and the Personalities*. London: HarperCollins.

Reid, Alastair and Henry Pelling. 2005. *A Short History of the Labour Party*. Basingstoke: Palgrave Macmillan.

Rowthorn, Bob. 1983. "The Past Strikes Back" in Stuart Hall and Martin Jacques (eds.) *The Politics of Thatcherism*. London: Lawrence & Wishart.

Schmidt, Vivien. 2008. "Discursive Institutionalism: The Explanatory Power of Ideas and Discourse". *Annual Review of Political Science* 11.

Seldon, Arthur (ed.). 1980. *Is Monetarism Enough? Essays in Refining and Reinforcing the Monetary Cure for Inflation*. IEA Readings n° 24. London: Institute of Economic Affairs.

Sennett, Richard. 2010. "Le New Labour n'a pas su comment gouverner". *Mediapart.fr*. 30 April.

Thatcher, Margaret. 1980. "Speech to Conservative Women's Conference". Festival Hall, London, 21 May.

—. 1981. "Press Conference for the British Press". Southern Cross Hotel, Melbourne, 7 October.

Taylor, Robert. 1978. *The Fifth Estate: Britain's Unions in the Modern World*. London: Pan Books.

Thompson, Noël. 2006. *Political Economy and the Labour Party: The Economics of Democratic Socialism, 1884-2005*. Oxford: Routledge.

Trades Union Congress. 1968. *Report of 100th Annual Trades Union Congress*. London: TUC.

—. 1970. *Report of 102nd Annual Trades Union Congress*. London: TUC.

—. 1975. *Report of 107th Annual Trades Union Congress*. London; TUC.

—. 1978. *Report of 110th Annual Trades Union Congress*. London: TUC.

Watt, David. 1968. *Financial Times*. 8 March.

Weber, Max. 1978. *Economy and Society: An Outline of Interpretive Sociology*. Vol. 1. Eds Guenther Roth and Claus Wittich. Berkeley and Los Angeles: University of California Press. [first publication in English 1968]

Wickham-Jones, Mark. 1996. *Economic Strategy and the Labour Party: Politics and Policy Making, 1970-1983*. Basingstoke: Macmillan.

Notes

[1] The authors thank the anonymous reviewer for their perceptive comments, which helped us clarify parts of our argument. The remaining confusions, of course, are entirely ours.

[2] We do not mean to suggest that Margaret Thatcher used the term "sound" as a strict synonym for "scientific". The expression "functional substitute", which we borrow loosely from Robert Merton (1968: 88), implies that a given word in a given discourse fulfills an analogous function to another word in another discourse. Our claim is therefore that the term "sound" denotes in Thacherite language the same empirical validity that academic language ascribes to "scientific" views. In her eyes, the policies she advocated were sound because they worked, just like scientific theories are considered valid because they work.

[3] This chapter only tangentially deals with the assumption by economics of an epistemic status akin to the natural sciences. Rather, we take economics as a science in its own right and investigate its relationship to political discourse.

[4] We borrow the term "common-sense constructs" from Stuart Hall's early work on ideology in the media (Hall, 1973: 86).

[5] What we offer here is nothing but a variation on the Kuhnian account of what happens during phases of normal science and crisis science (Kuhn 2012). Kuhnian insights into the mechanisms of political change were first systematically applied by Peter Hall (1990; 1992).

[6] It should be noted, however, that the intellectual and organizational history of neoliberalism goes all the way back to the 1930s and that it was never as circumstantial a project as the AES, which might help explain their differing fates.

[7] Hayek and Friedman both started out as statisticians. Hayek was the head of the Austrian Institute for Business Cycle Research, an institute collecting data on economic fluctuations which he founded in Vienna in 1927 and which is now the largest economic research institute in Austria. Friedman for his part began collecting data on income and wealth distribution after he joined the National Bureau of Economic Research in Washington in 1937 and was awarded his doctorate in 1946 on the basis of this work.

[8] For instance, at the 1975 TUC Congress, the Communist trade unionist Ken Gill used economic research to attack the policies pursued by the Labour government: "Professor Turner, the Cambridge professor, said in the July issue of *New Society*, 'Each stage of State intervention has moved through a similar sequence, virtually: remonstrance, confrontation, disintegration spread over three or four years. And since the 1950s, each such cycle has operated with a successively higher background of unemployment'" (Trades Union Congress, 1975: 463).

[9] "Congress is critical of the White Paper of the 11th July as an interference in free collective bargaining and a limitation by law on wage costs in the public sector. It believes that this merely treats the symptoms of inflation instead of tackling the underlying causes of instability in the UK economy. Congress, to maintain levels of employment, urges: (a) the introduction of selective import controls; (b) a public investment programme to raise productivity and create jobs; (c) strict control over the prices of essential items in working class family budgets; (d) the lowering of interest rates; (e) the control of export of capital; (f) the raising of consumption by increasing pensions, social insurance benefits, and social public expenditure. Congress urges trade union action to achieve: (i) a shorter working week and an extension of statutory and annual holidays; (ii) the reduction and elimination of overtime; (iii) a low pay target of £40 minimum basic rate" (Trades Union Congress 1975, 462).

[10] By "political discourse", we mean both the "coordinative discourse" through which ideas are circulated among policy actors (experts, technocrats, politicians, journalists, etc.) and the "communicative discourse" through which ideas are conveyed to the public. For the distinction between these two forms of discourse, see Schmidt (2008).

[11] More than reception, one should rather talk of appropriation, as historians and sociologists have stressed, to better emphasize the active role played by those "receiving" ideas.

[12] The exact quotation is the following: "The broad-based rejection of left policies by the public was based on an intuitive grasp of these inadequacies, rather than any explicit 'turning to the right' of the population" (Auberbach 2007, 28).

[13] The model offered here draws loosely on the literature on mobilisation, notably the paradigms of resource mobilisation and structure of opportunity (McAdam *et al.* 2001), as well as on the sociology of political crises (Dobry 1986).

BIBLIOGRAPHY

Achouche, Mehdi. 2012. "Biopolitique et biopouvoir au XXIème siècle: les biotechnologies dans le débat politique américain contemporain." *Revue Représentations*, February.

Adams, Samuel Hopkins. 1906. *The Great American Fraud, Articles on the Nostrum Evil and Quacks, in Two Series*, reprinted from *Collier's Weekly*. New York: P. F. Collier & Son.

Allen, Ann Taylor. 2000. "Feminism and Eugenics in Germany and Britain, 1900-1940: A Comparative Perspective", *German Studies Review*, Vol. 23, No. 3 (Oct., 2000), pp. 477-505.

"The American Medical Association called A Trust". 1910. *Journal of the American Medical Association* 54.23 (June 4). [1876]

American School of Magnetic Healing v. McAnnulty. 187 U.S. 94. Supreme Court of the US. 1902. Justia (n.d). To be consulted at: https://supreme.justia.com/cases/federal/us/187/94/. (accessed 3 Jan. 2015)

"A Bad Bunch." Collier's 6 May 1911, p. 10.

Amossy, Ruth. 2006. *L'argumentation dans le discours*. Paris: Armand Colin.

Appel T. A. 2010. "The Thomsonian Movement, the Regular Profession, and the State in Antebellum Connecticut: A Case Study of the Repeal of Early Medical Licensing Laws." *Journal of the History of Medicine and Allied Sciences* 65.2, pp. 153-86.

Arata, Stephen. 1996. "The Sedulous Ape: Atavism, Professionalism, and Stevenson's *Jekyll and Hyde*" in S. Arata, *Fictions of Loss in the Victorian Fin de Siècle*. Cambridge: Cambridge University Press, pp. 33-53.

Aristotle, *Politics*. 1916. London: J.M. Dent and Sons.

Aristote. *Rhétorique.* 1996. Paris : Grasset, Collection Le Livre de Poche [circa 329-323 BCE].

Arneson, Richard J.1999. "What, if Anything, Renders All Humans Morally Equal?" in Dale Jamieson (ed.). *Singer and His Critics*, Oxford: Blackwell, pp. 103-128.

Ashworth, John. 1987. *"Agrarians" and "Aristocrats": Party Political Ideology in the United States, 1837-1846.* Cambridge [Cambridgeshire], New York: Cambridge University Press.

Atkinson, Mabel. 1910. "The Feminist Movement and Eugenics", *The Sociological Review*, Volume a3, Issue 1, pages 51–56, January.

Auerbach, Paul. 2007. "The Left Intellectual Opposition in Britain, 1945-2000: The Case of the Alternative Economic Strategy" in Francis King (ed.) *Is There no Alternative? Historical Problems of Socialist Economic Strategies*. Socialist History Society Occasional Papers No. 24. London: Socialist History Society.

Benedetti, Gregory. 2014. "L'émergence d'une classe politique noire post-raciale, 2000-2012, le cas du nord-est des États-Unis: tensions, rupture, et continuité idéologique." PhD. Le Havre University.

Benn, Tony. 1975. *A New Course for Labour*. Nottingham: Spokesman.

Bentham, Jeremy. 2007. *An Introduction to the Principles of Morals and Legislation*. New York: Dover. [1780].

Berezow, Alex & Campbell, Hank. 2012. "Are Democrats Really the 'Pro-Science' Party?" Real Clear Politics, September 10. To be consulted at: http://www.realclearpolitics.com/articles/2012/09/10/are_democrats_really_the_pro-science_party_115367.html

Besson, Cyril. 2014. "Le glaive contre le caducée: de l'indétermination de la figure du médecin chez Robert Louis Stevenson" in Lise Dumasy-Queffélec and Hélène Spengler (eds.), *Médecine, sciences de la vie et littérature en France et en Europe de la Révolution à nos jours*. Genève: Droz, pp. 259-60.

"Beware, O Ye Doctors!" *New-York Tribune* 16 May 1910, p. 12.

Birke, Lynda. 1986. *Women, Feminism and Biology*. Brighton: Wheatsheaf Books Ltd.

Bland, Lucy. 1995. *Banishing the Beast: English Feminism and Sexual Morality, 1885-1914*. London: Penguin, 1995.

Blöde, Charles A. 2000. "Dr. F.J. Gall's System of the Functions of the Brain" in Nicholas J. Wade (ed.), *The Emergence of Neuroscience in the Nineteenth Century: Phrenology Examined*. London: Routledge/Thoemmes Press, pp. 1-88.

Blundell, John. 2007. "Achieving Change: What We Can Learn from Margaret Thatcher". Heritage Lecture n° 1000. Washington, DC: Heritage Foundation, December.

Blyth, Mark. 2002. *Great Transformations: Economic Ideas and Institutional Change in the 20th Century*. Cambridge: Cambridge University Press.

Boer, Deng, Richard Monastersky, Lauren Morello, Sara Reardon & Jeff Tollefson. 2015. "Obama budget seeks big boost for science." *Nature, News and Comments*, February 3. To be consulted at: http://www.scientificamerican.com/article/obama-s-budget-seeks-big-boost-for-science/

Bourdieu, Pierre. 1991. *Language and Symbolic Power* (translated from the French). Cambridge: Polity. [1982]

—. 2000. *Propos sur le champ politique.* Lyon: Presses Universitaires de Lyon.

—. 2013. *Distinction: A Social Critique of the Judgment of Taste* (translated from the French). London: Routledge. [1984]

Boyd, R., P. Gasper & J. D. Trout (eds). 1991. *The Philosophy of Science*, Cambridge, Massachusetts: MIT Press.

Brennan, Andrew and Lo, Yeuk-Sze. 2011. "Environmental Ethics," *The Stanford Encyclopedia of Philosophy,* Fall Edition. Edward N. Zalta (ed.). To be consulted at: http://plato.stanford.edu/archives/fall2011/entries/ethics-environmental/. (accessed 9 December 2014)

Briggs, Finney. 1914. "A the people want it." *Medical Freedom* 4 (Dec. 1914), p. 57.

Britt, Robert Roy. 2008. "Scientists Say Bush Stifles Science and Lets Global Leadership Slip." livescience.com, January 20th. To be consulted at: http://www.livescience.com/9574-scientists-bush-stifles-science-lets-global-leadership-slip.html

Britton, Andrew. 1991. *Macroeconomic Policy in Britain 1974-1987.* Cambridge: Cambridge University Press.

Brooke, John L. 2009. "Cultures of Nationalism, Movements of Reform, and the Composite-Federal Polity: From Revolutionary Settlement to Antebellum Crisis." *Journal of the Early Republic* 29.1 (2009), pp. 1-33.

Brown, Ian. 1988. "Who were the Eugenicists? A study of the formation on an early twentieth century pressure group", *History of education*, 17, pp. 295-307.

Brown, Timothy & White, John. 1837. *Messages of Gen. Andrew Jackson: With a Short Sketch of His Life.*

Buck v. Bell. 274 U.S. 200. Supreme Court of the US. 1927. *Supreme Court Collection.* Legal Information Inst. Cornell U Law School (n.d.) To be consulted at: http://www.law.cornell.edu/supremecourt/text/274/200. Accessed 2 Jan. 2015)

Burdett, Carolyn. 1998. "The Hidden Romance of Sexual Science: Eugenics, the Nation and the Making of Modern Feminism" in Lucy Bland and Laura Doan (eds.) *Sexology in Culture: Labelling Bodies and Desires*. Chicago: The University of Chicago Press.

—. 2002. "From *The New Werther* to numbers and arguments: Karl

Pearson's eugenics" in Roger Luckhurst and Josephine Mcdonagh (eds.), *Transactions and Encounters: Science and Culture in the Nineteenth Century*. Manchester: Manchester University Press, pp. 204-231.

Burke, Edmund. 1964. *Reflections on the Revolution in France*. London: Dent. [1790]

Bush, George W. "President Discusses Global Climate Change." To be consulted at: http://georgewbush-whitehouse.archives.gov/news/releases/2001/06/20010611-2.html

Bynum, W. F. 1994. *Science and the Practice of Medicine in the Nineteenth Century*. Cambridge [England]: New York: Cambridge University Press.

Callaghan, James. 1976. "Speech to the Labour Party Conference". Blackpool, 28 September.

Callicott, J. Baird. 2014. *Thinking like a Planet. The Land Ethic and the Earth Ethic*. New York: Oxford University Press.

Cambridge Political Economy Group. 1974. *Britain's Economic Crisis*. Nottingham: Spokesman.

Carnegie, Andrew. 1963. *Autobiography of Andrew Carnegie*. NY: Pyramid. [1920]

—. 1901. *The Gospel of Wealth and Other Timely Essays*. New York: Century.

Cartwright, Nancy. 1983. *How the Laws of Physics Lie*. London: Oxford University Press.

Cavalieri, Paola and Peter Singer (ed.). 1996. *The Great Ape Project. Equality Beyond Humanity.* New York: St Martin.

Chalmers, Alan F. 1999. *What is this thing called Science?* (third edition). Indianapolis: Hackett Publishing Company, Inc.

Chambers, Robert. 1994. *Vestiges of the Natural History of Creation and Other Evolutionary Writings*. Chicago: University of Chicago Press.

Charaudeau, Patrick. 2005. *Le discours politique. Les masques du pouvoir.* Paris: Vuibert.

—. & D. Maingueneau. 2002. *Dictionnaire d'analyse du discours.* Paris: Seuil.

Chassay, Jean-François. 2009. *Si la science m'était contée : des savants en littérature*. Paris: Editions du Seuil.

Cheneseau, Jean-Jacques. 1995. *Le langage du politique.* Poitiers: Méthodes et Stratégies.

Chiron, Yves. 1987. *Edmund Burke et la Révolution française*. Paris: Téqui.

Chronicling America. Historic American Newspapers. Library of Congress (n.d). To be consulted at: http://chroniclingamerica.loc.gov. (accessed 1 Aug. 2014)

Clark, Anna. 2007. "Gender and Radicalism" in Kelly Boyd and Rohan McWilliam (eds.), *The Victorian Studies Reader*. New York: Routledge, pp. 191-206.

Clarke, Edwin., Jacyna, L. S. 1987. *Nineteenth-Century Origins of Neuroscientific Concepts*. Berkeley: University of California Press.

Clarke, Peter. 1997. "The Keynesian Consensus and Its Enemies: The Argument over Macroeconomic Policy in Britain since the Second World War" in David Marquand and Anthony Seldon (eds), *The Ideas that Shaped Post-War Britain*. London: Fontana.

"The Closing of the League." *Medical Freedom* 6 (Sept. 1916), p. 2.

Coates, Ken and Tony Topham. 1972. *The New Unionism: The Case for Workers' Control*. London: Peter Owen.

Cockett, Richard. 1994. *Thinking the Unthinkable: Think-Tanks and the Economic Counter-Revolution, 1931-1983*. London: HarperCollins.

Col, Norbert. 2001. *Burke, le contrat social et les révolutions*. Rennes: PU Rennes.

Congressional Record 47 (6 July 1911), pp. 2659-75.

Coon, Deborah J. 1992. "Testing the Limits of Sense and Science: American Experimental Psychologists Combat Spiritualism, 1880–1920." *American Psychologist* 47.2: 143-51.

Cooter, Roger. 1984. *The Cultural Meaning of Popular Science: Phrenology and the Organization of Consent in Nineteenth-Century Britain*. Cambridge: Cambridge University Press.

Cosslett, Tess. 1992. *The "Scientific Movement" and Victorian Literature*. Brighton: The Harvester's Press Limited.

Coventry, Charles B. 1845. "History of Medical Legislation in the State of New York." *New York Journal of Medicine* 4, pp. 151-61.

Crackanthorpe, Montague. 1907. *Population and Progress*, London, Chapman & Hall, ltd.

CSE London Working Group. 1980. *The Alternative Economic Strategy: A Response by the Labour Movement to the Economic Crisis*. London: CSE.

Cunliffe, John. 2000. "Introduction: Left-Libertarianism—Historical origins." *The Origins of Left-libertarianism: An Anthology of Historical Writings*. Eds. Peter Vallentyne and Hillel Steiner. New York: Palgrave, pp. 1-20.

Cyrulnik, Boris, Elisabeth de Fontenay, Peter Singer. 2013. *Les Animaux aussi ont des droits*. Paris: Editions du Seuil. "The Daily Lesson." *Medical Freedom* 5 (Oct. 1915), p. 27.

Dale, Peter Allan. 1989. *In Pursuit of a Scientific Culture*. Madison: The University of Wisconsin Press.

Dardenne, Emilie. 2008a. "Les nouvelles frontières de l'altérité ? Peter Singer contre le spécisme," *L'autre*, Janine Dove-Rume, Michel Naumann, Tri Tran (eds.). Tours : Université François-Rabelais, pp. 469-488.

—. 2010a. "The Reception of Peter Singer's Theories in France". *Society and Animals. Journal of Human-Animal Studies* 18, vol. 2, pp. 205-218.

—. 2010b. "From Jeremy Bentham to Peter Singer," *Revue d'Etudes Benthamiennes* n°7, septembre, To be consulted at: http://etudes-benthamiennes.revues.org/204. Followed by a reply by Peter Singer: http://etudes-benthamiennes.revues.org/212.

Darwin, Charles. 1874. *The Descent of Man*. New York: American Publishers Corporation.

—. 1909. *On the Origin of Species by Means of Natural Selection, or the Preservation of Favoured Races in the Struggle for Life.* New York: P.F. Collier & Son.

Daston, Lorraine, and Fernando Vidal. 2004. *The Moral Authority of Nature*. Chicago: University of Chicago Press.

Davidoff, Leonore and Catherine Hall. 2002. *Family Fortunes: Men and Women of the English Middle Class 1780-1850*. Revised Edition. Milton Park: Routledge.

Davies, John D. 1955. *Phrenology Fad and Science*. New Haven: Yale University Press.

Davin, Anna. 1978. "Imperialism and Motherhood." *History Workshop Journal* 5, pp. 9-65.

Dawson, Michael. 2001. *Black Visions. The Roots of African-American Political Ideologies.* Chicago: The University of Chicago Press.

Delamont Sara & Lorna Duffin. 2013. *The Nineteenth-century Woman: Her Cultural and Physical World.* London: Routledge.

Delap, Lucy. 2004. "Feminist and anti-feminist encounters in Edwardian Britain", *Historical Research*, vol. 77, no. 196 (May).

Descartes, René. 1646. "Lettre au Marquis de Newcastle," 23 novembre 1646, *Oeuvres et lettres*. Paris: Gallimard, bibliothèque de la Pléiade, 1983, b, pp.1256-1257.

Desmond, Adrian. 2007. "Evolution Before Darwin" in Kelly Boyd and Rohan McWilliam (eds.), *The Victorian Studies Reader*. New York: Routledge, pp. 255-267.

Dhombres, Nicole. 1989. *Les Savants en Révolution, 1789-1799.* Calmann-Lévy: Cité des Sciences et de l'Industrie.

Dixon, Keith. 19998. *Les évangélistes du marché: Les intellectuels britanniques et le néo-libéralisme*. Paris: Raisons d'agir.

Dobry, Michel. 1986. *Sociologie des crises politiques*. Paris: Presses de la Fondation nationale des sciences politiques.

Donaldson, Sue and Will Kymlicka. 2011. *Zoopolis. A Political Theory of Animal Rights*. Oxford: Oxford University Press.

Dryden, Linda, Stephen Arata and Eric Massie (eds.). *Stevenson and Conrad: Writers of Land and Sea*. To be consulted at: researchrepository.napier.ac.uk/.../Introduction_15_ (accessed December 6, 2015).

"Dugald Stewart 1753-1828." International Association for Scottish Philosophy (n.d.) To be consulted at: http://www.scottishphilosophy.org/dugald-stewart.html. (accessed 31 March 2014)

Durant, John R. 1979. "Scientific naturalism and social reform in the thought of Alfred Russel Wallace", *The British Journal for the History of Science* Vol. 12, No. 1 (March), pp. 31-58.

Edward Francis, Mrs. 1911. "Race Suicide", *The Vote*, January 21, p. 152-53.

Egnal, Marc. 2009. *Clash of Extremes: The Economic Origins of the Civil War*. New York: Hill and Wang.

Ellis, Havelock. 1912. *The Task of Social Hygiene.* Boston: Houghton Mifflin.

"Eugenics". *Eugenics Review*, vol. 9, January-June 1912, p. 857.

Fairclough, Norman. 1989. *Language and Power.* London: Longman.

Farly Christopher and David Hodgson. 1972. *The life of Bertrand Russell*. Nottingham: Spokesman Books.

Fawcett, Milicent Garrett. 1890. "Home and Politics, An Address Delivered at Toynbee Hall". London: London Society for Women's Suffrage.

Fayol, Henri. 1918. "Administration industrielle et générale", *Bulletin de la Société de l'Industrie Minérale*, N° 10, 5-164, 1916. [reprints by Dunod after 1918]

Fernihough Anne. 2013. *Freewomen and Supermen: Edwardian Radicals and Literary Modernism*. Oxford University Press.

Ferry, Luc. 2000. "Des 'droits de l'homme' pour les grands singes? Non, mais des devoirs envers eux, sans nul doute". *Le Débat*, 108, février, pp. 163-167.

"Fight 'Doctors' Trust.'" *The New York Times* 10 January 1912.

Fisher, Irving. 1909. *National Vitality: its Wastes and Conservation*. Washington D.C.: Government Printing Office.

—. 1910a. "Defines Opposition to the Owen Bill." *The New York Times* 18 June: 8; reprinted as "The Layman's Reply to the Medical Freedom (?) League." Journal of the American Medical Association 54.5 (June 18), pp. 2133-34.

—. 1910b. "A Department of Dollars vs. a Department of Health." *McClure's Magazine* 35 (July), pp. 329-30.

—. 1912. *Memorial Relating to the Conservation of Human life as Contemplated by Bill (S. 1) Providing for a United States Public Health Service.* Washington D.C.: Government Printing Office.

—. 1913. "National Health and Medical Freedom. Both Sides of a Question of Public Interest." *Century Magazine* 85 (February), pp. 513-14.

Fisher, Misha. 2013. "The Republican Party isn't Really the Anti-Science Party." The Atlantic, November 11th. To be consulted at: http://www.theatlantic.com/politics/archive/2013/11/the-republican-party-isnt-really-the-anti-science-party/281219/

Flexner, Abraham. 1910. *Medical Education in the United States and Canada: A Report to the Carnegie Foundation for the Advancement of Teaching*. New York: Carnegie Foundation for the Advancement of Teaching.

Flower, Benjamin O. 1890a. "Notes on Living Problems of the Hour. Liberty of Citizenship Imperiled. A Consensus of Opinion on a Recent Notable Case. Introductory Statement by the Editor" *The Arena* 2 (July), pp. 229-30.

—. 1890b. "Praise for the Arena." *The Arena* 2 (November): n. pag.

—. 1890c. "A Transition Period." *The Arena* 3 (December), pp. 124-25.

—. 1891a. "An Object Lesson in Freedom." *The Arena* 3 (March), pp. 507-10.

—. 1891b. "Class Interests and the Rights of the People." *The Arena* 3 (March), pp. 510-11.

—. 1891c. "Herbert Spencer's Arraignment of Class Medical Laws." *The Arena* 3 (March), pp. 511-12.

—. 1898. "The Menace of Medical Monopoly" *The Arena* 9 (February): 400-16. Reprinted twice: "Restrictive Medical legislation and the Public Weal." *The Arena* 19 (June 1898), pp. 781-809, and "Medical Monopoly the Foe to Progress and Human Rights." *The Twentieth Century Magazine* 3 (Nov. 1910), pp. 148-56.

—. *The Century of Sir Thomas More*. 1896. Boston: The Arena Publishing Company.

—. 1902. "How Cleveland Stamped Out Smallpox" *The Arena* 27 (April), pp. 426-29.

—. 1907a. "The Recent Reckless and Irresponsible Attacks on Christian Science and its Founder, with a Survey of the Christian-Science Movement." *The Arena* 37 (January), pp. 47-67.
—. 1907b. "The Seven Alleged Delusions of the Founder of Christian Science Examined in the Light of History and Present-Day Research." *The Arena* 38 (October), pp. 408-22.
—. 1910a. *Christian Science as a Religious Belief and a Therapeutic Agent*. 2nd ed. Boston: Twentieth Century Company.
—. 1910b. "The Menace of a National Health Bureau." *The Twentieth Century Magazine* 2 (June), pp. 234-42.
—. 1910c. "Why We Oppose the Bills to establish a National Department or Bureau of Health." *The Twentieth Century Magazine* 2 (July), pp. 340-48.
—. 1910d. "Medical News Notes. The Fight for Medical Freedom." *The Twentieth Century Magazine* 2 (September), pp. 366-68.
—. 1910e. "Medical Monopoly the Foe to Progress and Human Rights." *The Twentieth Century Magazine* 3 (November), pp. 148-56.
—. 1910f. "Medical News Notes. The Revolt Against the Medical Hierarchy." *The Twentieth Century Magazine* 3 (December), pp. 271-73.
—. 1911a. "The Compulsory Medical Inspection of School Children." *The Twentieth Century Magazine* 4 (May), pp. 151-60.
—. 1911b. "The Battle for Medical Freedom." *The Twentieth Century Magazine* 4 (May), pp. 177-79.
—. 1911c. "The Chicago Conference." *Medical Freedom* 1 (December), pp. 2-3.
—. 1912a. "Try to Stop Free Speech!" *Medical Freedom* 1 (February), p. 2.
—. 1912b. "The Second Annual Meeting." *Medical Freedom* 1 (June), p. 3.
—. 1912c. "B.O. Flower Answers Irving Fisher." *Medical Freedom* 2 (September), p. 11.
—. 1912d. "An Appeal." *Medical Freedom* 2 (October), pp. 2-3.
—. 1913a. "National Health and Medical Freedom. Both Sides of a Question of Public Interest." *Century Magazine* 85 (February), pp. 512-13.
—. 1913b. "The Second General Conference of the League. Address of Welcome by B.O. Flower." *Medical Freedom* 2 (August), pp. 8-10.
—. 1914a. *Progressive Men, Women, and Movements of the Past Twenty-five Years*. Boston: The New Arena.
—. 1914b. "Position of Political Doctors is Fundamentally Unsound."

Medical Freedom 3 (July), p. 7.
—. 1914c. "An Earnest Word to Friends of Medical Freedom in Grave Crisis." *Medical Freedom*, 4 (October), pp. 25-26.
—. 1914d. "Rising Revolt Against the Burdens of State Medicine." *Medical Freedom* 4 (December), p.5.
—. 1915. "A Bird's-Eye View of the National league for Medical Freedom." *Medical Freedom* 4 (April), p. 117.
—. 1916. *Story of the Menace Trial: A brief sketch of this historic case with reports of the masterly addresses by Hon. J.L. McNatt and Hon. J.I. Sheppard, attorneys for the defense*. Aurora, MO: United States Publishing Co.
Flower Richard C. 1888. *The Allopathic Czar*. Boston: Massachusetts Medical Liberty League.
Foner, Eric. 1994. "The Meaning of Freedom in the Age of Emancipation." *Journal of American History* 81. 2 (September), pp. 435-60.
—. 1999. *The Story of American Freedom*. New York: W.W. Norton.
Fontenay, Elisabeth de. 2000. "Pourquoi les animaux n'auraient-ils pas droit à un droit des animaux ?" *Le Débat*, 108, février, pp. 138-155.
Forester, Tom. 1979. "Neutralising the Industrial Strategy" in Ken Coates (ed.), *What Went Wrong*. Nottingham: Spokesman.
Foucault, Michel. 1990. *The History of Sexuality, Vol. 1: An Introduction*. (translated from the French). New York: Vintage books. [1976]
Fowler, L.N. 1974. "The Principles of Phrenology and Physiology Applied to Man's Social Relations" in Charles Rosenberg and Carroll Smith-Rosenberg (eds), *Sex, Marriage and Society*. New York: Arno Press, pp. 5-135.
Friedman, Milton and Rose. 1980. *Free to Choose. A Personal Statement*. New York: Avon Books.
Fuchs, Michel. 1996. *Edmund Burke, Ireland and the Fashioning of the Self*. Oxford: Voltaire Foundation.
Gall, Franz Josef. 1835. *On the Functions of the Brain and of Each of Its Parts: With Observations on the Possibility of Determining the Instincts, Propensities, and Talents, or the Moral and Intellectual Dispositions of Men and Animals, by the Configuration of the Brain and Head*. Boston: Marsh, Capen & Lyon.
Galton, Francis. 1865. *Hereditary Talent and Character.* Originally published in Macmillan's Magazine, 12, pp. 157-166. available at http://www.cimm. jcu.edu.au/ hist/stats/galton/macindex.htm
—. 1883. *Inquiries into Human Faculty and its Development*. London: Macmillan.

—. 1892. *Hereditary Genius: An Inquiry into Its Laws and Consequences.* 2nd ed. London: Macmillan.[1869]
—. 1908. *Memories of my Life*. London: Methuen & Co.
—. 1909. *Essays in Eugenics.* London: Eugenics Education Society.
Ganzin, Michel. 1969. *La pensée politique d'Edmund Burke*. Thèse soutenue à l'Université d'Aix-Marseille, février.
Garner, Les. 1984. *Stepping Stones to Women's Liberty: Feminist Ideas in the Women's Suffrage Movement, 1900-1918*. Fairleigh Dickinson University Press.
Gay, Peter. 1969. *The Enlightenment: The Science of Freedom*. New York: W.W. Norton & Company, Inc., 1969.
Gillispie, Charles Coulston. 1959. "Science in The French Revolution" in *Proceedings of the National Academy of Sciences of the United States of America*, Princeton University, May, vol. 45(5): 677–684.
—. 2004. *Science and Polity in France: The Revolutionary and Napoleonic Years*. Princeton: Princeton University Press.
Gingras, Yves. 2010. *Propos sur les sciences, entretiens avec Yanick Villedieu*. Paris: Raisons d'agir.
Glyn, Andrew and Bob Sutcliffe. 1972. *British Capitalism, Workers and the Profits Squeeze*. Harmondsworth: Penguin.
Golan, Tal. 2004. *Laws of Men and Laws of Nature the History of Scientific Expert Testimony in England and America*. Cambridge, Mass.: Harvard University Press.
Gould, Stephen Jay. 1981. *The Mismeasure of Man*. New York: W.W. Norton & Company, Inc.
Grossman LA. 2013. "The Origins of American Health Libertarianism." *Yale Journal of Health Policy, Law, and Ethics* 13.1, pp. 76-134.
Haldane, Charlotte. 1927. *Motherhood and Its Enemies*. London: Chatto and Windus.
Hall, Benjamin F. 1848. *The Trial of William Freeman for the Murder of John G. Van Nest : Including the Evidence and the Arguments of Counsel, with the Decision of the Supreme Court Granting a New Trial, and an Account of the Death of the Prisoner, and of the Post-Mortem Examination of His Body by Amariah Brigham, M.D., and Others*. Auburn: Derby, Miller & Co.
Hall, Lesley. 2004. "Hauling Down the Doubles Standard: Feminism, Social Purity and Sexual Science in Late Nineteenth Century Britain", *Gender & History*, Vol. 16, N°1, April, pp. 36-56.
Hall, Peter. 1990. "Policy Paradigms, Experts, and the State: The Case of Macroeconomic Policy-Making in Britain" in Stephen Brooks and Alain-G. Gagnon (eds.) *Social Scientists, Policy, and the State*. New

York: Praeger.

—. 1992. "The Movement from Keynesianism to Monetarism: Institutional Analysis and British Economic Policy in the 1970s". *Structuring Politics: Historical Institutionalism in Comparative Analysis* in eds. Sven Steinmo, Kathleen Thelen and Frank Longstreth. *Cambridge Studies in Comparative Politics*. Cambridge: Cambridge University Press.

Hall, Stuart. 1973. "A World at One with Itself" in Stanley Cohen and Jock Young (eds.) *The Manufacture of News: Social Problems, Deviance and the Mass Media*. London: Constable.

Hamilton, Alexander or Madison, James. 1788. "The Federalist Papers: No. 49. Method of Guarding Against the Encroachments of Any One Department of Government by Appealing to the People Through a Convention." *The New York Packet*. Feb. 5. The Federalist Papers. Library of Congress (n.d.). To be consulted at: http://thomas.loc.gov/home/histdox/fed_49.html. (accessed 2 Jan. 2015)

Hamilton, Cicely. 1912a. "Women in the Great State", in H.G. Wells et al., *Socialism and the Great State: Essays in Construction*. New York: Harper and Brother, pp. 219-248.

—. 1912b. In H.G. Wells et al., *Socialism and the Great State: Essays in Construction*. New York: Harper and Brother, p. 226.

Handelsman, Jo. 2015. "Obama Administration Aims to Erase Science Talent Deficit." *Scientific American*, March 23rd. To be consulted at: http://www.scientificamerican.com/article/obama-administration-aims-to-erase-science-talent-deficit/?print=true.

Harris, R.W. 1963. *Political Ideas, 1760-1792*. London: Gollancz.

Harvie, Christopher. 1981. "The Politics of Stevenson," in Jenni Calder (ed.), *Stevenson and Victorian Scotland*. Edinburgh: Edinburgh University Press, pp. 107-125.

Hatch, Nathan O. 1989. *The Democratization of American Christianity*. New Haven: Yale University Press.

Hatch, Robert A. "The Scientific Revolution, Definition – Concept – History" at http://users.clas.ufl.edu/ufhatch/pages/03-Sci-Rev/SCI-REV-Teaching/03sr-definition-concept.htm (accessed Sept. 28, 2015).

Hausman, Daniel, ed. 2007. *The Philosophy of Economics: An Anthology* (3rd ed.). New York: Cambridge University Press, 2007. [1984]

Haywood, Andrew. 2007. *Political Ideologies*. New York: Palgrave Macmillan (4th edition). [1992].

Hazen, Craig James. 2000. *The Village Enlightenment in America: Popular Religion and Science in the Nineteenth Century*. Urbana: University of Illinois Press.

Henry, John. 2002; *The Scientific Revolution and the Origins of Modern Science*. London: Palgrave Macmillan. [1997]

Henson, Louise, Geoffrey Cantor et al. 2004. *Culture and science in the nineteenth-century media*. Aldershot: Ashgate.

Hickson, Kevin. 2005. *The IMF Crisis of 1976 and British Politics*. London: Tauris Academic Studies.

Historical Highlights. U.S. Department of Health & Human Services, n.d. To be consulted at: (accessed 3 Jan. 2015).

Hobsbawm, Eric. 1969. *Industry and Empire*. London: Weidenfeld & Nicolson.

Hobson, J. A. 1910. *A Modern Outlook*. London: Herbert & Daniel.

Hofstadter, Richard. 1955. *The Age of Reform : from Bryan to F.D.R*. New York: Vintage Books.

Holland Stuart. 1975. *The Socialist Challenge*. London: Quartet.

—. 1978. *Beyond Capitalist Planning*. Oxford: Blackwell.

Hollinger, David A. 1989. "The Problem of Pragmatism in American History." *In the American Province: Studies in the History and Historiography of Ideas*. Baltimore: Johns Hopkins UP, pp. 23-43.

Holly, W. 1989. "Credibility and Political Language", in Ruth Wodak (ed.), *Language Power and Ideology*. Amsterdam: John Benjamins, pp. 115-35.

Holt, Michael F. 1999. *The Rise and Fall of the American Whig Party*: Jacksonian Politics and the Onset of the Civil War. New York: Oxford University Press.

Holton, Richard and Rae Langton. 1999. "Empathy and Animal Ethics," in Dale Jamieson (ed.), *Singer and His Critics*. Oxford: Blackwell, pp. 209-232.

Howe, Daniel Walker. 1997. *Making the American Self: Jonathan Edwards to Abraham Lincoln*. Cambridge, Mass.: Harvard University Press.

Hugins, Walter Edward. 1960. *Jacksonian Democracy and the Working Class, a Study of the New York Workingmen's Movement, 1829-1837*. Stanford: Stanford University Press.

Hume, David. 2007. *A Treatise of Human Nature*, David Fate Norton and Mary J. Norton (eds.), 2 vols. Oxford: Oxford University Press. [1739]

"Imaginary Dangers Proclaimed." *The New York Times* 27 Sept. 1910: 12.

"Investing in America's Future. Barack Obama and Joe Biden's Plan for Science and Innovation." To be consulted at: http://pubs.acs.org/cen/news/pdf/FactSheetScience.pdf. 2008.

Jacobson v. Commonwealth of Massachusetts. 197 U.S. 11. Supreme Court of the US. 1905. Justia (n.d). To be consulted at:

https://supreme.justia.com/cases/federal/us/197/11/. (accessed 3 Jan 2015)

James, William. 1987. "Review of 'Rapport sur les progrès et la marche de la physiologie générale en France' by Claude Bernard (1868)." *Essays, Comments, and Reviews*. Cambridge, MA: Harvard UP.

Jamieson, Dale (ed.). 1999a. *Singer and His Critics*. Oxford: Blackwell.

—. 1999b. "Singer and the Practical Ethics Movement," in Dale Jamieson (ed.), *Singer and His Critics*. Oxford: Blackwell, 1999, p. 1-17.

Jeffreys, Sheila. 1985. *The Spinster and Her Enemies - Feminism and Sexuality 1800-1930.* London: Pandora.

Johnston, Robert D. (ed.). 2004. *The Politics of Healing: Histories of Alternative Medicine in Twentieth-Century North America*. New York: Routledge.

—. 2003. *The Radical Middle Class: Populist Democracy and the Question of Capitalism in Progressive Era.* Portland, Oregon. Princeton: Princeton UP.

Jolly, Roslyn. 2009. *Robert Louis Stevenson in the Pacific: Travel, Empire and the Author's Profession.* Farnham, England: Ashgate Publishing Ltd.

Jones, Greta. 1995. "Women and Eugenics in Britain: the case of Mary Scharlieb, Elizabeth Sloan Chesser and Stella Brown", *Annals of Science*, 51, pp. 481-502.

—. 1980. *Social Darwinism and English social thought: the interaction between biological and social theory*, Brighton: Harvester, 1980.

Jordan, Thomas Edward. 1992. *The Degeneracy Crisis and Victorian Youth*, State University of New York Press.

Jordanova, Ludmilla. 1994. "Melancholy Reflection: Constructing an Identity for Unveilers of Nature", in Stephen Bann (ed.), *Frankenstein, Creation and Monstrosity*. London, Reaktion Books.

Kavanagh, Dennis. 2004. "Conclusion: Reading and Misreading Old Labour" in Anthony Seldon and Kevin Hickson (eds.) *New Labour, Old Labour: The Wilson and Callaghan Governments, 1974-1979.* London: Routledge.

Keiper, Adam. 2009. "Science and the Obama Administration." The New Atlantis, Winter. To be consulted at: http://www.thenewatlantis.com/publications/science-and-the-obama-administration

Kett, Joseph F. 1968. *The Formation of the American Medical Profession; the Role of Institutions, 1780-1860*. New Haven: Yale University Press.

Kilpatrick, Andrew and Tony Lawson. 1980. "On the Nature of the Industrial Decline in the UK". *Cambridge Journal of Economics* 4:1.

King, Anthony and Robert Wybrow. 2001. *British Political Opinion, 1937-*

2000: The Gallup Polls. London: Politico's.
Kinealy, Arabella. 2011. *Feminism and Sex Extinction*. London: Fisher & Unwin.
Knopf, Sigard Adolphus. 1910. "The Owen Bill for the Establishment of a Federal Department of Health, and its Opponents." *Popular Science Monthly* 77 (October), pp. 373-78.
Kolbert, Elizabeth. 2015. "The G.O.P.'s War on Science Gets Worse." The New Yorker, May 6th 2015. To be consulted at: http://www.newyorker.com/news/daily-comment/gop-war-on-science-gets-worse.
Krantz, Susan Lufkin. 2002. *Refuting Peter Singer's Ethical Theory. The Importance of Human Dignity*. Westport: Praeger.
Kuhn, Thomas S. 1957. *The Copernican Revolution: Planetary Astronomy in the Development of Western Thought*. Harvard: Harvard University Press.
—. 1962. *The Structure of Scientific Revolutions*, Chicago: The University of Chicago Press.
Laplanche, Jean et Jean-Bertrand Pontalis. 1968. *Vocabulaire de la psychanalyse*. Presses univ. de France. [Translated into English as The language of psychoanalysis. Karnac Books, 1988]
Latour, Bruno. 1989. *La science en action*. Paris: La Découverte.
Le Bart, Christian. 1998. *Le discours politique*. Paris: Hachette.
—. 2003. "L'analyse du discours politique : de la théorie des champs à la sociologie de la grandeur". *Mots. La ville, entre dire et faire*, 72 (July).
Lepeltier, T. (ed.). 2013. *Histoire et philosophie des sciences*. Auxerre: Sciences Humaines Editions.
Levere, Trevor H. 1980. "Coleridge and Human Science" in Marsha P. Hanen, Margaret J. Osler and Robert G. Weyant (eds.), *Science, Pseudo-Science and Society*. Waterloo: Wilfrid Laurier University Press, pp. 171-192.
Lewis, George Cornewall. 1849. *On the Influence of Authority in Matters of Opinion*. London: Savill and Edwards Printers.
"Liberty." *Collier's* 3 June 1911, p. 9.
Liptak, Adam. 2014. "Supreme Court Rejects Contraceptives Mandate for Some Corporations. Justices Rule in Favor of Hobby Lobby" *The New York Times* 30 June. To be consulted at: http://www.nytimes.com/2014/07/01/us/hobby-lobby-case-supreme-court-contraception.html. Accessed 3 Jan 2015.
Love, Rosaleen. 1979. "Alice in Eugenics-Land': Feminism and Eugenics in the scientific careers of Alice Lee and Ethel Elderton", *Annals of Science*, Volume 36, Issue 2, 1979.

Lucas, John. 1971. "Conservatism and Revolution in the 1880's" in John Lucas (ed.), *Literature and Politics in the Nineteenth Century*. London: Methuen and Co. LTD, pp. 173-219.

Ludlam, Steve. 1990. "Labourism and the Disintegration of the Postwar Consensus: Disunited Trade Union Economic Policy Responses to Public Expenditure Cuts, 1974-1979". PhD thesis, University of Sheffield.

Lyotard, Jean-François. 1979. *La Condition postmoderne, Rapport sur le Savoir*. Paris : Les Editions de Minuit.

McAdam, Doug, Sidney Tarrow and Charles Tilly. 2001. *Dynamics of Contention*. Cambridge: Cambridge University Press.

MacDuffie, Allen. 2006. "Irreversible Transformations: Robert Louis Stevenson's *Dr. Jekyll and Mr. Hyde* and Scottish Energy Science", in *Representations* 96, Autumn, pp. 1-20.

Makepeace, Clare. 2009. "To What Extent was the Relationship Between Feminists and the Eugenics Movement a 'Marriage of Convenience' in the Interwar Years?". *Journal of International Women's Studies*, 11(3), September, pp. 66-80.

Malakoff, David. 2015. "Science Gets its Moment in Obama's 2015 State of the Union." Science Mag, January 26th. To be consulted at: http://news.sciencemag.org/funding/2015/01/science-gets-its-moment-obamas-2015-state-union

Marin-Lamellet, Jean-Louis. 2014. "What is the cocoon but a dark cabinet?': Benjamin O. Flower, Print Culture and the Legitimisation of Fringe Science in the 1890s." *Mémoires du livre / Studies in Book Culture* 6.1 (fall 2014): n. pag. To be consulted at : http://www.erudit.org/revue/memoires/2014/v6/n1/1027697ar.html. (accessed 3 Jan. 2015)

Maxwell, Clerk. 1890. "Introductory Lecture on Experimental Physics." *Scientific Papers*. Ed. W.D. Nivens. New York: Dover Publications.

Medical Freedom 1 (Sept. 1911), p. 2.

—. 2 (Oct. 1912), p. 2.

"Medical Freedom." Editorial. *The New York Times* 18 May 1910, p. 10.

Merton, Robert. 1968. *Social Theory and Social Structure.* New York: The Free Press, 1968 (3rd ed.). [1949]

Minkin, Lewis and Patrick Seyd in William Paterson and Alastair Thomas (eds.) 1977. "The British Labour Party". *Social Democratic Parties in Europe*. London: Croom Helm.

Mohr, James C. 1993. *Doctors and the Law: Medical Jurisprudence in Nineteenth-Century America.* New York: Oxford University Press.

Mullins, Jeffrey A. 1995. "Duties to Science, Duties to God : Medical

Theory, Physiology, and the Discourse on Morality in Nineteenth-Century America." *Transactions and Studies of the College of Physicians of Philadelphia* 17, pp. 45-60.

Murray Leslie, R. 1911. "Woman's Progress in relation to eugenics", *Eugenic Review*, January; 2(4): 282–298, pp. 283, 291-292.

Nairn, Tom. 1979. "The Future of Britain's Crisis". *New Left Review* 13/14.

National League for Medical Freedom (The). 1910. *First Report of The National League for Medical Freedom Issued to its Members*. August. New York: National League for Medical Freedom.

"A New Combination Against the American Medical Association." *Journal of the American Medical Association* 54.22 (May 28, 1910), p. 1791.

Newmarch, Eugenia. 1911. "Women and the Race", *The Englishwoman*, April, pp. 33-40.

Newspaper.com. *Historical Newspapers from 1700s-2000s.* (n.d). To be consulted at: http://www.newspapers.com. (accessed 1 Aug. 2014)

Noll, Mark A. 1985. "Common Sense Traditions and American Evangelical Thought" *American Quarterly* 37.2 (1 July), pp. 216-38.

Novak, William J. 1996. *The People's Welfare : Law and Regulation in Nineteenth-Century America*. Chapel Hill, N.C.: University of North Carolina Press.

Obama, Barack. "Inaugural Address." January 21, 2009. To be consulted at: https://www.whitehouse.gov/blog/2009/01/21/president-barack-obamas-inaugural-address

—. "Remarks by the President at the National Academy of Sciences Annual Meeting." April 27, 2009. To be consulted at: https://www.whitehouse.gov/the-press-office/remarks-president-national-academy-sciences-annual-meeting

—. "Remarks by the President on the "Educate to Innovate" Campaign." November 23, 2009. To be consulted at: https://www.whitehouse.gov/the-press-office/remarks-president-education-innovate-campaign

—. "Remarks by the President at the White House Science Fair." October 18, 2010. To be consulted at: https://www.whitehouse.gov/the-press-office/2010/10/18/remarks-president-white-house-science-fair

—. "Remarks by the President at the White House Science Fair." February 7, 2012. To be consulted at: https://www.whitehouse.gov/photos-and-video/video/2012/02/07/president-obama-speaks-white-house-science-fair#transcript

—. "Remarks by the President at the White House Science Fair." April 22,

2013. To be consulted at: https://www.whitehouse.gov/the-press-office/2013/04/22/remarks-president-2013-white-house-science-fair

—. "Remarks by the President at the White House Science Fair." May 27, 2014. To beconsulted at: https://www.whitehouse.gov/the-press-office/2014/05/27/remarks-president-white-house-science-fair

—. "State of the Union Address." January 20, 2015. To be onsulted at: https://www.whitehouse.gov/the-press-office/2015/01/20/remarks-president-state-union-address-january-20-2015.

—. "Remarks by the President at the White House Science Fair." March 23, 2015. To be consulted at https://www.whitehouse.gov/the-press-office/2015/03/23/remarks-president-white-house-science-fair

Orwell, George. 1946. *Politics and the English Language*, First published in *Horizon*, London, April. To be consulted at: http://www.orwell.ru/library/essays/politics/english/e_polit/

"The Organization of Ill Health." Journal of the American Medical Association 55.5 (July 30, 1910): 421.

Pandora, Katherine. 2009. "Popular Science in National and Transnational Perspective: Suggestions from the American Context." Isis 100, pp. 346-58.

Pankhurst, Christabel. 1913. *The Great Scourge and how to end it, 1913.* London: Pankhurst.

—. 1914. "Concerning Damaged Goods", *The Suffragette*, 20 February.

Parsons, Wayne. 1990. *The Power of the Financial Press: Journalism and Economic Opinion in Britain and America*. New Brunswick, NJ: Rutgers University Press. [1989][

Passeron, Irène. 2008. "La République des sciences : réseau des correspondances, des académies et des livres scientifiques au 18ème siècle." Revue du Dix-huitième siècle, pp. 5-28. Available at: https://hal.archives-ouvertes.fr/hal-00361446v2

Pearson, Karl. 1888. *The Ethics of Free Thought*. London: T.F. Unwin.

—. 1892. *The Grammar of Science*. London: Walter Scott.

—. 1894. "Woman and Labour", *Fortnightly Review* (1894), pp. 561-77.

—. *The Life, Letters and Labours of Francis Galton*. Cambridge: Cambridge University press, 1914-30.

—. 2013. "The Woman's Question", 1885, in *The Ethic of Freethought and Other Addresses and Essays*. London: Forgotten Books. [1901]

Peirce, Charles Sanders. 1935. "Issues of Pragmaticism." The Monist 15.4 (Oct. 1905): 481-99. Collected Papers of Charles Sanders Peirce. Volumes V and VI: Pragmatism and Pragmaticism and Scientific Metaphysics. Eds. Charles Hartshorne, Paul Weiss. Cambridge, MA: Harvard UP, pp. 438–63.

Pethick-Lawrence, Emmeline. 1908. "What the vote means for those who are fighting the battle", *Votes for Women*, January, p. 49
Petrina, Stephen. 2008. "Medical Liberty: Drugless Healers Confront Allopathic Doctors, 1910–1931." *Journal of Medical Humanities* 29.4 (Aug. 7), pp. 205–230.
Pew Research Center. 2009. "Public Praises Science: Scientists Fault Public." July 9th. To be consulted at: http://www.people-press.org/2009/07/09/public-praises-science-scientists-fault-public-media/
Phillipson, Nicholas. 2010. *Adam Smith: An Enlightened Life*. London: Penguin.
Pielke, Roger Jr. & Roberta Klein. 2009. "The Rise and Fall of the Science Advisor to the President of the United States." Springer Science+Business Media B.V. 2009. To be consulted at: http://sciencepolicy.colorado.edu/admin/publication_files/resource-2719-2009.05.pdf
Popper, Karl. 2002. *The Logic of Scientific Discovery*. London, New York: Routledge Classics. [1935]
Postman, Neil. 1992. *Technopoly*. New York: Vintage Books.
"Power of the Progressive Economics: the Clinton Years." Center for American Progress. October 28, 2011. To be consulted at: https://www.americanprogress.org/issues/economy/report/2011/10/28/10405/power-of-progressive-economics-the-clinton-years/
Pugh, Martin. 1978. *Electoral Reform in Peace and War 1906-18*, London: Routledge.
Rabban, David M. 1992. "The Free Speech League, the ACLU, and Changing Conceptions of Free Speech in American History." *Stanford Law Review* 45.1 (Nov. 1), pp. 47-114.
Ranelagh, John.1991. *Thatcher's People: An Insider's Account of the Politics, the Power, and the Personalities*. London: HarperCollins.
Raynaud, Dominique. 2015. Scientific Controversies: A Socio-Historical Perspective on the Advancement of Science (translated from the French). London: Transaction Publishers. [2003]
Reid, Alastair and Henry Pelling. 2005. *A Short History of the Labour Party*. Basingstoke: Palgrave Macmillan.
Reid, Charles. "Ethos, Pathos, and Logos: A Guide for Scientists." To be consulted at: http://charlesreid1.com/wordpress/2013/01/ethos-pathos-and-logos-a-guide-for-scientists (accessed Aug. 14, *2015).*
Reid, Julia. 2006. Robert Louis Stevenson, Science, and the Fin de Siècle. Basingstokes: Palgrave Macmillan.
Report of the Inter-departmental Committee on Physical Deterioration, Vol. I, Report and appendix, London: His Majesty's Stationery Office,

1904, available on https://archive.org/details/b21358916 [consulted 12/08/20115].

Richardson, Angelique. 2008. *Love and Eugenics in the Late Nineteenth Century: Rational Reproduction and the New Woman*. Oxford: OUP.

—. 2008. "The life sciences", in David Bradshaw, *A Concise Companion to Modernism*, John Wiley & Sons, pp. 6-33, p. 20

—. 2001. "'People talk a lot of nonsense about Heredity': Mona Caird and anti-eugenic feminism", in *The New Woman in Fiction and in Fact: Fin de Siècle Feminisms*. London: Palgrave.

Robb, George. 1997. 'Race Motherhood: Moral Eugenics vs Progressive Eugenics, 1880-1920', in Claudia Nelson and Ann Holmes (eds), *Maternal Instincts: Images of Motherhood and Sexuality in Britain, 1875-1925*. London: Macmillan Press.

—. 1998. 'Eugenics, Spirituality, and Sex Differentiation in Edwardian England: the Case of Frances Swiney', *Journal of Women's History* 10.3: 97- 117.

—. 1996. 'The Way of All Flesh: Eugenics, Degeneration and the Gospel of Free Love'. *Journal of the History of Sexuality* 6: 589-603.

Rolston III, Holmes. 1994. "Value in Nature and the Nature of Value," in *Philosophy and the Natural Environment*. Robin Attfield, Andrew Belsey (eds.). Cambridge: Cambridge University Press, pp. 13-30.

—. 1999. "Respect for Life: Counting what Singer Finds of no Account," in Dale Jamieson (ed.), *Singer and His Critics*. Oxford: Blackwell, pp. 247-268.

Roosevelt, Theodore. 1908. "Eighth Annual Message." 8 Dec. Gerhard Peters and John T. Woolley. The American Presidency Project (n.d.) To be consulted at: http://www.presidency.ucsb.edu/ws/?pid=29549. (accessed 2 Jan. 2015).

Rosenfeld, Sophia A. 2011. *Common Sense: A Political History*. Cambridge, MA: Harvard UP.

Rostow, Walt Whitman. 1960. The Stages of Economic Growth: A non-Communist Manifesto. Cambridge: CUP.

Rothbard, Murray N. 1976. "The Methodology of Austrian Economics" in Edwin G. Dolan, ed. The Foundations of Modern Austrian Economics. Kansas City, Sheed and Ward, 1976, b75-88. To be consulted at: http://oll.libertyfund.org/titles/104 (accessed 22/05/2015).

Rowold, Katharina. 2009. The Educated Woman: Minds, Bodies, and Women's Higher Education in Britain, Germany, and Spain, 1865-1914. New York and London: Routledge.

Rowthorn, Bob. 1983. "The Past Strikes Back" in Stuart Hall and Martin

Jacques (eds.) *The Politics of Thatcherism*. London: Lawrence & Wishart.

Saleeby, Caleb William. 1910. "Racial Poison", *Eugenics Review* 2 (April 1910- January 1911), pp. 30-52.

—. 1909. *Parenthood and Race Culture: An Outline of Eugenics*. New York: Moffat, Yard.

—. 1911. *Woman and Womanhood: A Search for Principles*. New York and London: Mitchell Kennerley.

Scharlieb, Mary. 1912. *Womanhood and Race-Regeneration*. New York: Moffat, Yard.

Schmidt, Vivien. 2008. "Discursive Institutionalism: The Explanatory Power of Ideas and Discourse". *Annual Review of Political Science* 11.

Schumpeter, Joseph, Alois. 2011. The Theory of Economic Development: An Inquiry into Profits, Capital, Credit Interst and the Business Cycle, Transaction (translated from the German in 1934). [German edition 1911]

Scott, Joan Wallach. 1999. *Gender and the Politics of History*. New York: Columbia University Press.

Searle, G.R. 1976. *Eugenics and politics in Britain, 1900-14* Leyden: Noordhoff International Publishing.

Secord, James A. 1994. Introduction to Robert Chambers, *Vestiges of the Natural History of Creation and Other Evolutionary Writings*. Chicago: University of Chicago Press.

—. 2007. "Domesticating Evolution" in Kelly Boyd and Rohan McWilliam (eds.), *The Victorian Studies Reader*. New York: Routledge, pp. 268-281.

Seldon, Arthur (ed.). 1980. *Is Monetarism Enough? Essays in Refining and Reinforcing the Monetary Cure for Inflation*. IEA Readings n°24. London: Institute of Economic Affairs.

Sennett, Richard. 2010. "Le New Labour n'a pas su comment gouverner". *Mediapart.fr*. 30 April.

Shapiro, Michael J. 1981. *Language and Political understanding: the Politics of Discursive Practices*. New Haven: Yale University Press.

Shelley, Mary. 2012. *Frankenstein*. London: Penguin English Library. [1818]

Singer, Peter. 1973. "The Triviality of the Debate Over 'Is-Ought' and the Definition of 'Moral'". American Philosophical Quarterly, X, 1, January, pp. 51-56.

—. 1995. *Animal Liberation*, Second Edition with a New Preface by the Author. London: Pimlico [First Edition 1975, Second Edition Jonathan Cape, 1990].

—. 2011. *Practical Ethics*, Third Edition. Cambridge: Cambridge University Press [First Edition, 1980, Second Edition, 1993].

—. 1999. "A Response," in Dale Jamieson (ed.), *Singer and His Critics*, Oxford: Blackwell, pp. 269-335.

—. 2010, "The Great Debate—Can Science Tell Us Right from Wrong?," talk given at Arizona State University on November 6. To be consulted at: http://thesciencenetwork.org/programs/the-great-debate/peter-singer. Accessed 21 October 2014.

Snow, Charles. 1964. *The Two Cultures and A Second Look*. Cambridge: Cambridge University Press.

Soloway, Richard A. 1982a. *Birth control and the population question in England 1877-1930*. Chapel Hill.

—. 1982b. "Feminism, Fertility and Eugenics in Victorian and Edwardian England" in George Lachmann, *Political Symbolism in Modern Europe,*Brunswick: Transaction, pp. 121-145.

—. 1990. *Demography and Degeneration, Eugenics and the Declining Birthrate in Twentieth Century Britain*. The University of North Carolina Press.

Sombart, Werner. 2013. *Krieg und Kapitalismus*. BoD. [1913]

Speedy, Andrew. 2003. "Global Production and Consumption of Animal Source Foods," *The Journal of Nutrition*, November 1, vol. 133, n° 11, p. 4048S.

Spencer, Herbert. "Over-Legislation". 1992. *The Westminster Review* for July, 1853. Reprinted in *The Man Versus The State, with Six Essays on Government, Society, and Freedom*. Econlib site, p. 7.42 To be consulted at:
http://www.econlib.org/library/LFBooks/Spencer/spnMvS5.html (accessed 22/05/2015).

Spencer, Ivor D. 1941. "William L. Marcy: 'An Educated Northern Democrat'." *New York History* 22.2, pp. 180-90.

Stevenson, D.T. 1911. "The Eugenic Vote", *The Common Cause*, 31 August 1911, p. 356.

Stevenson, Robert Louis. 1915. *Across The Plains With Other Memories And Essays*. London: Chatto &Windus.

—. 1925. *In the South Seas*. New York, Charles Scribner's Sons.

—. 1892. *A Footnote to History: Eight Years of Trouble in Samoa*. London: Cassell Company Limited.

—. 1901. "A plea for Gas Lamps" in *Virginibus Puerisque and Other Papers*. London: Chatto & Windus, pp. 271-8. [1878]

Stiles Anne. 2006. "Robert Louis Stevenson's *Jekyll and Hyde* and the Double Brain" in *SEL* 46, 4, Autumn, pp. 879-900.

Stolberg, Sheryl Gay. 2009. "Obama Puts His Own Spin on the Mix of Science With Politics." *The New York Times*. March 10th. To be consulted at: http://query.nytimes.com/gst/fullpage.html?res=9C07E0DB173EF933A25750C0A96F9C8B63.

Sussman, Herbert. 1968. *Victorians and the Machine, the Literary Response to Technology*. Cambridge, Mass.: Harvard University Press.

Swift, Jonathan. 1973. *The Writings of Jonathan Swift*. New York: Norton.

Taft William H. 1909. "First Annual Message." 7 Dec. Gerhard Peters and John T. Woolley. *The American Presidency Project* (n.d.) To be consulted at: http://www.presidency.ucsb.edu/ws/?pid=29550. (accessed 2 Jan. 2015)

—. 1910. "Second Annual Message." 6 Dec. Gerhard Peters and John T. Woolley. *The American Presidency Project* (n.d.) To be consulted at: http://www.presidency.ucsb.edu/ws/?pid=29551. (accessed 2 Jan. 2015)

Taylor, Frederik Winslow. 1911. *The Principles of Scientific Management*. New York, Harpers and Brothers.

Taylor, Robert. 1978. *The Fifth Estate: Britain's Unions in the Modern World*. London: Pan Books.

Thatcher, Margaret. 1980. "Speech to Conservative Women's Conference". Festival Hall, London, 21 May.

—. 1981. "Press Conference for the British Press". Southern Cross Hotel, Melbourne, 7 October.

Thierry, Patrick. 2010. *Burke, le futur en héritage*. Paris: Michalon.

Thompson, Noël. 2006. *Political Economy and the Labour Party: The Economics of Democratic Socialism, 1884-2005*. Oxford: Routledge.

Tooley, Sarah A. 1896. "The Woman Question: An Interview with Madame Sarah Grand", *Humanitarian*, 8, p. 168.

Torode, B. 1991. *Review of Fairclough, N., Language and Power, Discourse and Society* (2) (1), pp.121-2.

Trades Union Congress. 1968. *Report of 100th Annual Trades Union Congress*. London: TUC.

—. 1970. *Report of 102nd Annual Trades Union Congress*. London: TUC.

—. 1975. *Report of 107th Annual Trades Union Congress*. London; TUC.

—. 1978. *Report of 110th Annual Trades Union Congress*. London: TUC.

United States House of Representatives Committee on Government Reform. Minority Staff Special Division Investigation. "Politics and Science in the Bush Administration." August 2003. To be consulted at:

http://socrates.berkeley.edu/~scotch/science_policy/Bush_Admin.pdf

Vallentyne, Peter. 2000. “Left-Libertarianism: A Primer.” *Left Libertarianism and Its Critics: The Contemporary Debate*. Eds. Peter Vallentyne and Hillel Steiner. New York: Palgrave, pp. 1-20.

Van de Vyver, Gisèle, et Jacques Reisse. 1991. *Les Savants et la politique à la fin du XVIIIe siècle*. Bruxelles: Editions de l'Université de Bruxelles.

Veblen, Thorstein. 1978. *The Theory of Business Enterprise*. Transaction Publishers. [1899]

—. *The theory of the leisure class; an economic study of institutions*. 2005. Aakar Books.

—. 1898. “Why is economics not an evolutionary science?” *The Quarterly Journal of Economics*, 1898, vol. 12, no 4, pp. 373-397.

Votes for Women. 21 January 1909, p. 208; 5 August 1913, p. 9.

Wallace, Alfred Russel. 1913. *Social Environment and Moral Progress*, New York: Cassell and Company.

Watt, David. 1968. *Financial Times*. 8 March.

Weber, Max. 1978. *Economy and Society: An Outline of Interpretive Sociology*. Vol. 1. Eds Guenther Roth and Claus Wittich. Berkeley and Los Angeles: University of California Press. [first publication in English 1968]

Whale, John ed. 2000. *Edmund Burke's Reflections on the Revolution in France*: New Interdisciplinary Essays. Manchester: Manchester UP.

—. *Science and Polity in France: The Revolutionary and Napoleonic Years*. 2004. Princeton: Princeton University Press.

White, Ian. 2010. “No vote – no census: an account of some of the events of 1910–1911” *Office for National Statistics,* Population Trends nr 142, Winter.

“Who is Right? *Medical Freedom* 1 (March 12), p. 10.

Wickham-Jones, Mark. 1996. *Economic Strategy and the Labour Party: Politics and Policy Making, 1970-1983*. Basingstoke: Macmillan.

Wiebe, Robert. 1967. *The Search for Order, 1877-1920*. New York: Hill and Wang.

Willrich, Michael. 2011. *Pox: An American History*. New York: The Penguin Press.

Wilson, John. 2001. “Political discourse”, in Deborah Schiffrin, Deborah Tannen, Heidi E. Hamilton, *The Handbook of Discourse Analysis*, Oxford: Blackwell, pp. 398-414.

Wintrhop Evans, Mrs. 1911. “The Mother as a Factor in Human Progress”, *The Vote*, September 9, 1911, pp. 248-49.

Witemeyer, Hugh. 1979. *George Eliot and the Visual Arts*. New Haven: Yale University Press.

"Woman and Natural Selection". Interview With Dr. Alfred Russel Wallace, *The Daily Chronicle*, 4 December 1893. To be consulted at: http://people.wku.edu/charles.smith/wallace/S736.htm.

Yeo, Richard. 2001. *Science in the Public Sphere*. Burlington: Ashgate Publishing Company.

Zunz, Olivier. 2000. *Why the American Century?* Chicago: University of Chicago Press.

CONTRIBUTORS

PAUL AUERBACH received his PhD from the University of Wisconsin, USA and is Reader in Economics at Kingston University, UK. His most recent published work is the book *Socialist Optimism: An Alternative Political Economy for the Twenty-First Century*. His research interests include the measurement of economic growth, the economics of competition, and the relationship between education and economic development.

GREGORY BENEDETTI is a senior lecturer in American studies at Grenoble Alpes University (France). He completed a doctoral thesis on "The emergence of a post-racial black political class in the north-east of the United States, 2000-2012." His publications include "Barack Obama's Political Discourse: Black and More than Black" (2012) and "The Post-Racial Mystery: the Advent of a New Racial Dialectics in American Politics?" (2012).

CYRIL BESSON is a senior lecturer in Scottish studies at Grenoble Alpes University (France). A specialist of nineteenth-century Scottish literature, he has worked on the concepts of nation and nationhood, authority and the various modes of "naturalization" of the arbitrariness of social and cultural constructs. He co-edited the issue of the electronic journal *Representation* (CEMRA) on "History and Political Discourse" with Pr. V. Molinari, and the book, *Ville et environnement: regards croisés sur le monde postcolonial* (2014) with André Dodeman.

GILLES CHRISTOPH teaches at the École Normale Supérieure de Lyon (France). His doctoral thesis was dedicated to retracing the intellectual history of Anglo-American neoliberalism over the course of the twentieth century. His research now centers on the relationship between science and philosophy in the wake of the science wars of the 1990s.

ÉMILIE DARDENNE is a senior lecturer in English at Rennes 2 University (France). She has published articles on Victorian and contemporary animal ethics, utilitarianism, the works of Henry Stephens Salt and of Peter Singer. She has organised several international conferences on these topics. She is the co-editor of *Deux siècles d'utilitarisme*, with Malik Bozzo-Rey (2011) and is currently working on a volume entitled *La libération animale, quarante ans plus tard* with Valéry Giroux and Enrique Utria (forthcoming 2016).

JULIE HOMCHICK (PhD, University of Washington) is a Lecturer in the Department of Communication at Seattle University whose work primary focuses on the public understanding of science and scientific controversies. She has presented her research for regional, national and international organizations and has published in various outlets, including *POROI: An Interdisciplinary Journal of Rhetorical Analysis and Invention*, *Science and Education* and *Keywords and Controversies in the Rhetoric of Science and Technology*. Her most recent works have looked at the creationist movement in the United States, museum representations of science, and she is currently finished a co-authored piece on the Paleo Diet.

CHRISTIAN LEBLOND is a former student of the École Normale Supérieure de Fontenay-St Cloud. After the Agrégation d'Anglais in 1985 he taught in Madrid before taking up a position at Pierre-Mendès-France-Grenoble II University (Department of Economics). He is now a senior lecturer in American history at Grenoble Alpes University (France). He defended his doctoral dissertation in Geneva in 1999 on the North American Free Trade Agreement, and joined the ILCEA research center, where his main focus is the history of economic concepts and inter-cultural issues in business.

MARC LENORMAND is a senior lecturer in British Studies at Paul-Valéry University in Montpellier, France. In 2012, he completed a doctoral thesis on "British trade unions in a context of crisis of Labour, extension of industrial disputes and rise of the New Right: a critical history of the 1978-1979 'Winter of Discontent'", and has published a series of articles about the history of the British trade union movement, women in trade unions, trade union democracy, labour politics, and the Conservative anti-union agenda. His research now focuses on the intellectual history of the labour movement in Britain in the second half of the 20th century.

JEAN-LOUIS MARIN-LAMELLET is a PhD candidate at Lumière University in Lyon (France). His research interests are Gilded Age and Progressive Era America, American cultural and intellectual history and the history of the book and journalism. His PhD research focuses on the intellectual biography of Boston reform editor Benjamin Orange Flower (1858-1918). His recent publications include "'Acquainting America with work in foreign experiment stations': Benjamin Orange Flower and *The Arena*, 1889-1909." *(2014);*"'What is the cocoon but a dark cabinet?': Benjamin O. Flower, Print Culture and the Legitimisation of Fringe Science in the 1890s" (2014) and Benjamin O. Flower's "War on War" (2015).

VÉRONIQUE MOLINARI is Professor of British studies at Grenoble Alpes University (France) where she teaches British history. Her research focuses on the women's movement in Edwardian and Interwar Britain as well as on women's participation in politics. She is co-editor with Donna Spalding-Andreolle of *Women and Science, 17th Century to Present: Pioneers, Activists and Protagonists (*2011) and the author of several books and articles including "'Schools of their own' - The Ladies' Medical College and London School of Medicine for Women" (2011) and "Educating and Mobilizing the New Voter: Interwar Handbooks and Female Citizenship in Great-Britain, 1918-1931" (2014).

JEFFREY MULLINS received his A.B. from Reed College and his Ph.D. from the Johns Hopkins University. He has been awarded faculty fellowships at the University of Oklahoma (History of Science) and Emory University (Center for Humanistic Inquiry), and is currently Professor of History at St. Cloud State University, where he is Director of Graduate Studies. His prior publications include "'Fitted to Receive the Word of God': Emotions and Scientific Naturalism in the Religious Revivals of the 1830s" in the *International Social Science Review*, and his article "'Hurrah for Hanging': Monsters, Irony, and the Contested Meaning of Horror in Nineteenth-Century America" is forthcoming in the journal *Interdisciplinary Humanities.*

SARAH PEYROUX-ROBLIN is a senior lecturer at Franche-Comté University. A former student at ENS Cachan, she is an agrégée in English, and completed a thesis entitled "Les marginaux dans la poésie anglaise, 1770-1810." Her research mainly focuses on English literature during the French Revolution, especially on works of an anti-Jacobin sensibility. Her recent publications include "Robert Southey under Attack: Anti-Jacobin Propaganda and Literary Revisionism" (2011) and "Résonances de l'*Essay on Man* dans la poésie anti-jacobine".

INDEX